Multi-field Coupled Constitutive Model for Soils Based on Granular Thermodynamics Theory

Guangchang Yang

Multi-field Coupled Constitutive Model for Soils Based on Granular Thermodynamics Theory

Springer

Guangchang Yang
School of Future Cities
University of Science and Technology Beijing
Beijing, China

ISBN 978-981-95-0248-6 ISBN 978-981-95-0249-3 (eBook)
https://doi.org/10.1007/978-981-95-0249-3

This Springer imprint is published by the registered company Springer Nature Singapore Pte Ltd.
The registered company address is: 152 Beach Road, #21-01/04 Gateway East, Singapore 189721, Singapore

Contents

Chapter 1
Introduction

1.1 Background and Research Significance of the Topic Selection

The multi-field coupling theory of geotechnical media is a product of the intersection and permeation of disciplines such as geotechnical mechanics, seepage mechanics, heat transfer, and engineering technology. The research on this problem initially began in the water conservancy and hydropower industries, and now it has spread to fields such as geothermal energy, natural gas hydrate extraction, and nuclear waste storage. Geotechnical media are products naturally generated over a long period of geological time, and are porous media composed of different mineral combinations. It is influenced by various factors in nature, almost all of which interact in the form of fields and constantly change with the changes in time and space. The mechanical, physical, and chemical properties of rock and soil comprehensively reflect the deformation, failure, and interaction between the three-phase components of solid, liquid, and gas under external stress, seepage field, temperature field, chemical field, and other fields, namely the coupling effect. This coupling effect is often unidirectional or bidirectional and produces irreversible effects, such as fluid velocity and pressure, stress and deformation, temperature gradients, and processes between fluid solid chemical reactions. As shown in Fig. 1.1, the arrows between the fields indicate the coupling relationship, with their direction pointing toward the direction of action.

Due to the diversity and complexity of geotechnical systems, when establishing multi-field coupled mathematical models for geotechnical systems, three modeling methods are generally used based on specific situations: mechanism analysis, mixed analysis, and system analysis. Mechanism analysis method is to analyze the mechanism of interaction between multiple fields, using mechanical and mathematical methods to establish constitutive equations and solution conditions, and form a multi-field coupled distributed parameter mathematical model. Due to the need to

G. Yang, *Multi-field Coupled Constitutive Model for Soils Based on Granular Thermodynamics Theory*, https://doi.org/10.1007/978-981-95-0249-3_1

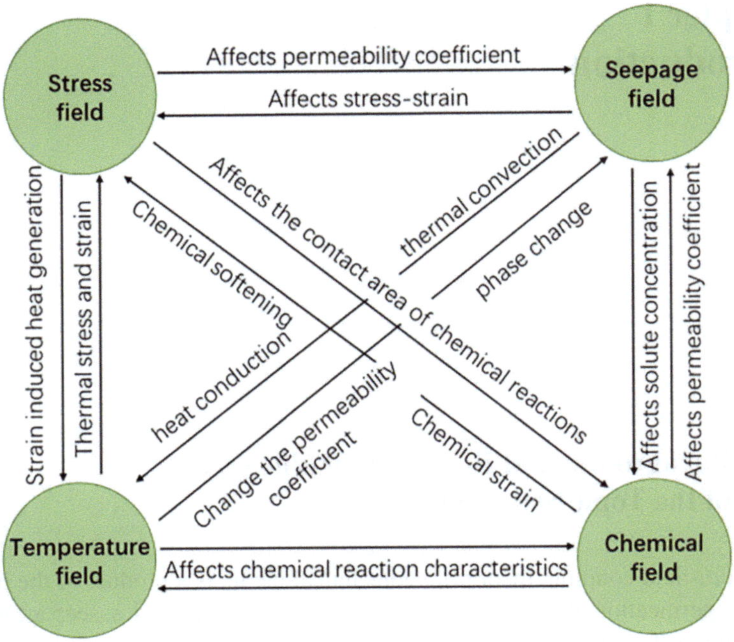

Fig. 1.1 Common fields and their coupling relationships in engineering geology

simplify complex problems during theoretical modeling, it is necessary to propose reasonable simplification assumptions for the studied object. However, these assumptions may not necessarily correspond to the actual situation, and coupled with the uncertainty of certain mechanical mechanisms in actual rock and soil systems, this method is generally used for mathematical modeling of simple systems. For complex multi-field coupling problems in rock and soil systems, this modeling method has significant limitations. The mixed analysis method is based on the constitutive equation obtained from mechanism analysis, combined with the coupling test analysis method, to establish a mathematical model of multi-field coupling of rock and soil systems. For complex coupling problems that cannot be expressed in simple mathematical expressions, empirical or semi-empirical relationships derived from experimental results can be used to describe them. This method is currently the most widely used method for establishing geotechnical coupling models, but its biggest drawback is the lack of rigorous and scientific theoretical basis. System analysis method is a method of estimating the mathematical model of a geotechnical system based on the corresponding output of a numerical simulation system under human input, with necessary mathematical calculations and data processing. The mathematical model established by this method is a lumped parameter model, often only targeting a specific problem, making it difficult to establish a widely applicable multi-field coupled mathematical model through this method.

The multi-field coupling problem of rock and soil mass is mainly studied based on the rock and soil mass, its occurrence environment, and engineering effects. By combining experimental and theoretical research, qualitative and quantitative analysis methods are used to study the multi-field coupling mechanism and coupling model, the numerical calculation method of the multi-field coupling model, and the deformation and failure mechanism of rock and soil mass under multi-field coupling. In terms of the application of multi-field coupling models in geotechnical media, the main focus is on the application research of hydraulic coupling models, such as in the fields of earth rock dam engineering, slope engineering, etc. Other multi-field coupling models have gradually been applied in many fields such as geothermal development, waste treatment, and energy infrastructure engineering. However, the guiding role of multi-field coupling analysis theory and methods in engineering construction is relatively limited, and there are very few that can be practically applied. Currently, most research is focused on the coupling effect of stress field and seepage field, and there is relatively little research on the three field coupling of thermo-hydro-mechanical (THM), which is an urgent need to be filled in the multi-field coupling theory.

As mentioned above, rock and soil media are in a certain geological environment and are influenced by various factors such as geostress, pore water, temperature, and chemical environment. These factors are interrelated and mutually constrained, forming a multi-field coupling effect of stress field, seepage field, temperature field, and chemical field in rock and soil mass. With the progress of society and the development of the national economy, there is an urgent need to solve the engineering problems of multi-field coupling of rock and soil mass mentioned above, and new multi-field coupling problems will continue to emerge with the resolution of relevant engineering problems. The study of multi-field coupling of rock and soil is the theoretical basis for these engineering problems. Its development can not only provide technical support for this, but also make strategic and breakthrough contributions. At the same time, it enriches the theoretical research of multi-field coupling of rock and soil, and has high academic value. Applying it to practical engineering will also generate huge social and economic benefits. Therefore, the multi-field coupling problem of rock and soil mass is a very necessary and valuable research topic, which can more objectively and truly reflect the essential properties of rock and soil media as the geological background of human engineering activities.

1.2 Research Status

For soil, it can be divided into two categories based on whether there is gas present in the soil: saturated soil and unsaturated soil. Wheeler [1] classified unsaturated soils into three types based on their saturation levels and the forms of water and gas: (1) At higher saturation levels (generally believed to be $S_r > 85\%$), the water phase is connected, while the gas phase exists in the form of isolated and closed bubbles, as shown in Fig. 1.2a. (2) At lower saturation levels (generally considered to be $S_r < 40\%$), the gas phase is connected while the water phase is not, as shown in Fig. 1.2b. (3) When the saturation is between the two, both the gas and water phases

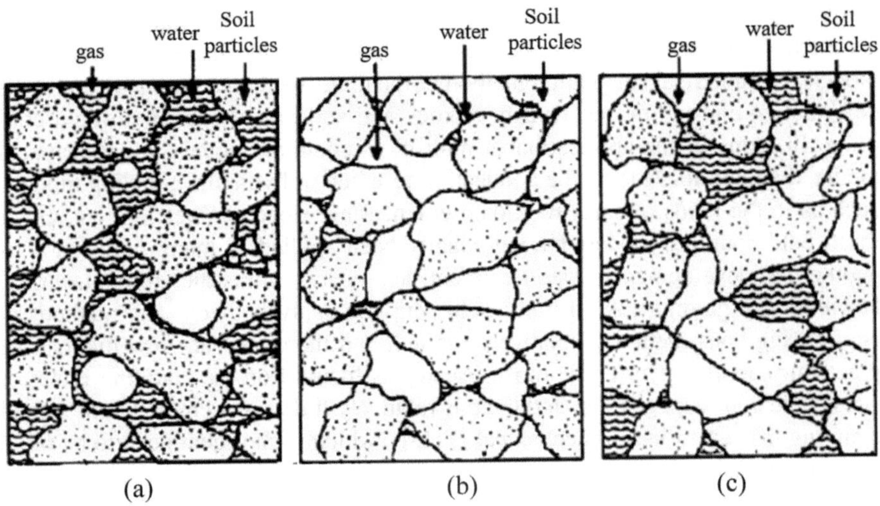

Fig. 1.2 Three types of unsaturated soils [1]: (**a**) Continuous water, discontinuous gas; (**b**) Continuous gas, discontinuous water; (**c**) Continuous water, continuous gas

are connected, as shown in Fig. 1.2c. For unsaturated soil containing sealed air, it is also known as gas bearing soil. Generally speaking, unsaturated soil refers to soil that is continuous in the gas phase and connected to the atmosphere. Since the pressure of the gas phase is the same as atmospheric pressure, the influence of gas phase pressure is generally not considered. However, for gas containing soil, due to the closed state of the internal gas, external disturbances can cause changes in gas volume and pressure, so the impact of internal gas changes on the soil needs to be considered. Sometimes, this type of soil is considered as a special form of soil independent of unsaturated soil [2].

The research process of unsaturated soil can be roughly divided into three stages: (1) Combining some research theories of saturated soil, attempting to provide an effective stress expression for unsaturated soil, such as Bishop's effective stress formula [3]; (2) The first International Conference on Unsaturated Soil Mechanics held in 1995 marked the beginning of the gradual improvement of unsaturated soils. So far, research on unsaturated soils has become increasingly in-depth, involving a wider range of soil types, research content, and methods, greatly enriching the theoretical and experimental foundations of unsaturated soils.

In terms of multi-field coupling analysis of soil, foreign research started earlier and has done a lot of work in the coupling analysis of mechanical field (M), temperature field (T), and seepage field (H) [4–6]. However, due to the complexity and variability of rock and soil media, there are still many shortcomings in research, and it is far from reaching a perfect level. Although the research on multi-field coupling of rock and soil has only a few decades of history, great progress has been made in the application of water conservancy, hydropower, geothermal, mining engineering, and other fields. The multi-field coupling we refer to generally includes the coupling between temperature and mechanical fields (TM), temperature and seepage field (TH), mechanical and seepage field (HM), and temperature-mechanical-seepage field (THM). In recent years, the influence of chemical field (C) has gradually attracted the attention of scholars, and the study of THMC four field coupling has also emerged. This chapter summarizes the models and experimental studies of several common coupling forms of saturated and unsaturated soils, as well as relevant research on gas bearing soils.

1.2.1 Hydromechanical (HM) Coupling Model of Soil

Hydraulic coupling refers to the coupling of the seepage field and stress field in the soil, and the interaction between the two is mainly reflected in two aspects: (1) Changes in the stress state in the soil will affect the soil structure, thereby changing the permeability characteristics of the soil, manifested as changes in the soil seepage field, which can be represented by empirical and theoretical relationships. (2) The changes in the seepage field in the soil will cause changes in the seepage force, thereby altering the original stress state in the soil, which is generally reflected through the principle of effective stress. The research on hydraulic coupling of soil mainly focuses on unsaturated soil, including experimental and constitutive model studies.

In the past few decades, although many constitutive models for unsaturated soils have been proposed [8, 9, 14–18], there are still many controversies over some basic issues, such as the selection of stress and strain state variables, and the theoretical framework of hydraulic coupling models. Alonso et al. [8] first established a constitutive model for unsaturated soil using dual stress variables of net stress and matric suction. Many scholars [10, 11, 14, 19] have borrowed this method to describe the hydraulic coupling characteristics of unsaturated soil. However, some scholars [7, 20, 21] have found that the dual stress variables of net stress and matric suction cannot describe some basic hydraulic coupling characteristics of unsaturated soils. Wheeler et al. [9] used Bishop's effective stress and modified suction as stress-strain variables to simulate the effects of saturation and volumetric strain on the hydraulic coupling of unsaturated soils, but this model is only applicable to isotropic loading conditions. Sheng et al. [14] and Sun et al. [15] extended Wheeler's model to anisotropic loading conditions, and the resulting model focuses more on the hydraulic characteristics of the stress-strain relationship, without highlighting the influence of deformation on hydraulic characteristics. In recent years, Muraleetharan et al. [22] have further improved the theoretical framework of the Wheeler model by using particle stress tensor and matric suction as dual stress variables. Although this model can simulate the dry-wet cycle characteristics well, there is still room for discussion regarding the simulation of loading and unloading cycles. Ghasemzadeh and Amiri [23] proposed a model that can simulate the hydraulic coupling characteristics of unsaturated soils by improving the Muraleetharan model. This model can consider the influence of volumetric moisture content on mechanical properties, such as irreversible compression during drying and changes in moisture content during loading and unloading. Most of the above models are only applicable to isotropic stress states. Liu and Muraleetharan [24, 25] proposed a model based on Muraleetharan's model framework that can be applied to multiple stress paths. This model can simulate the mechanical properties of unsaturated soil under complex loading paths well. However, under the condition of fixed main dry/main wet curves, the model cannot simulate the influence of plastic deformation on hydraulic properties, and the model is only applicable to unsaturated silt and sand.

In addition, some scholars [14, 16, 26–29] have established a hydraulic coupling constitutive model based on thermodynamic principles by analyzing the thermodynamic potential of unsaturated soil solid-liquid gas phases, as well as the dissipated energy during capillary hysteresis and skeleton deformation processes. Among them, Li [16] proposed a thermodynamic model framework for unsaturated soils based on the work, energy, and dissipation relationships of the solid-liquid gas three-phase system.

The accurate characterization of soil water characteristic lines is crucial for describing the hydraulic coupling problem of unsaturated soil, so the study of soil water characteristic curve models considering deformation effects is particularly important. One method is to establish a relationship between the inlet value parameters of the soil water characteristic curve and the volume deformation and porosity ratio, in order to reflect the influence of deformation on the soil water characteristic curve. Huang et al. [12] proposed a soil water characteristic curve model that can reflect deformation characteristics based on the Brooks Corcy model using this method. Gallipoli [30] assumed a power function relationship between the intake value and the porosity ratio, and based on the vG model [31], provided a soil water characteristic curve model considering deformation effects and hysteresis characteristics. Assouline [32] proposed a soil water characteristic curve model considering deformation characteristics by establishing the relationship between pore distribution parameters, air intake values, and soil dry density. In the above models, the influence of deformation on the soil water characteristic curve is reflected by establishing a relationship between the intake value and the porosity ratio. Another method is to establish a model based on experimental results and theoretical analysis. Sun et al. [33] established a soil water characteristic curve model considering hysteresis characteristics and changes in porosity based on experimental data. Tarantino [13] found through experimental research that there is a power function relationship between moisture content and suction, and established a soil water characteristic curve model considering changes in porosity based on the vG model. Zhou et al. [34] considered the deformation constraints of solid liquid gas three-phase mixtures and proposed a soil water characteristic curve model that can take into account the initial porosity ratio.

1.2.2 Thermomechanical (TM) Coupling Model of Soil

The coupling effect between temperature field and stress field in soil is mainly manifested in the following ways: (1) When the temperature field inside the soil changes, the change in temperature environment affects the physical and mechanical properties of the soil. (2) Temperature changes can generate thermal stress, leading to changes in the original stress field environment of the soil, and this in turn will lead to changes in the thermal characteristics of the soil and the generation of internal heat dissipation, resulting in a corresponding secondary adjustment of the temperature field of the soil. The coupling effect between temperature field and stress field affects and depends on each other until reaching a dynamic equilibrium state.

Hueckel and Baldi [39] first proposed an elastic-plastic thermodynamic constitutive model for saturated soil by extending the modified Cambridge model. The yield surface at different temperatures can be represented by the following equation:

$$f=\left(\frac{2p'}{p'_c}-1\right)^2+\left(\frac{2q}{Mp'_c}\right)^2-1=0 \tag{1.1}$$

where, p'_c is the maximum pre-consolidation effective stress, which controls the size of the yield surface. As the temperature increases, the yield surface will shrink. Using LY (load yield) curve to describe the variation of yield surface with temperature.

$$p'_c=p_{atm}\left(\frac{e_1-(1-\alpha_0\Delta T)\left[e_g-(1+e_0)\varepsilon_v^{tp}\right]}{\lambda-K_T}\right) \tag{1.2}$$

where, λ and K_T are the isothermal elastic-plastic modulus and elastic modulus, respectively; e_0 and e_1 represents the initial porosity ratio and the porosity ratio in the reference state, respectively; α_0 is the corresponding temperature influence parameter.

Due to the influence of temperature on the elastic range, the criteria for distinguishing loading and unloading differ from traditional elastic-plastic theory, which is not only related to stress paths but also to changes in temperature. Based on this theoretical framework, some scholars [35, 38, 39] have also proposed thermal coupling models for saturated soils, and these improved models further enhance the ability of the models to describe thermal characteristics. Cui et al. [35] proposed a thermal yield (TY) surface considering thermal yield based on thermoplastic theory and some experimental research results. It forms an elastic zone in the *T*-*p*′ plane with the LY yield surface, as shown in Fig. 1.3. The TY yield surface is related to the maximum temperature that the soil has been subjected to. Due to its introduction, the model can better simulate the thermal coupling characteristics of soil. Compared with the model proposed by Hueckel and Baldi [39], it has the following improvements: heating normally consolidated soil under constant pressure can

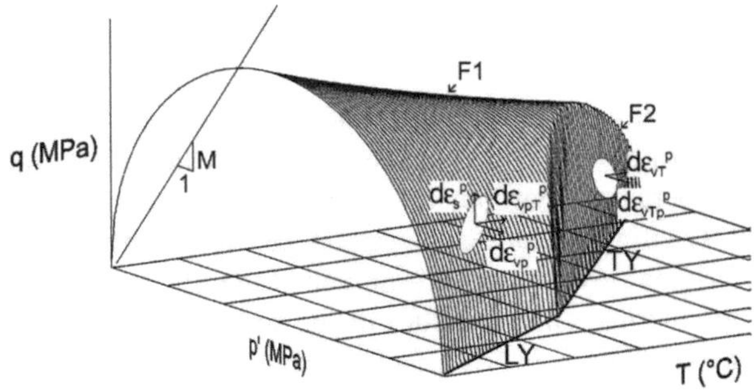

Fig. 1.3 Yield surfaces in p'-q-T spaces [35]

cause overconsolidation effect. Heating of overconsolidated soil will result in plastic deformation. But as the yield surface increases, the model parameters also increase accordingly, making the solution more complex.

Abule Naga et al. [38] proposed a thermomechanical coupling model under critical conditions, and the expression for the yield surface is as follows:

$$\frac{p'}{p'_c} = \frac{M^2 + \alpha^2}{M^2 + (\eta - \alpha)^2} \tag{1.3}$$

where, α is a parameter related to temperature. This expression can reflect the influence of temperature on the yield surface, as shown in Fig. 1.3, and improve the model's ability to simulate shear characteristics.

In addition, He et al. [40] studied the changes in temperature, pore pressure, and thermal stress around a spherical cavity in an infinitely large saturated porous medium using non-equilibrium thermal porous elasticity theory, and obtained asymptotic solutions for temperature, pore pressure, and thermal stress through Laplace transform. Hong et al. [41] compared three well-known thermodynamic models [35, 36, 42] and evaluated the strengths and weaknesses of each model, particularly in terms of its ability to simulate thermal strain. Zhou and Ng [43] proposed a thermomechanical coupling model for saturated soil based on the boundary surface plasticity theory, considering the influence of temperature on the size and shape of the boundary interface, which can simulate the characteristic of reduced shear modulus under small deformation.

The study of the thermal coupling characteristics of unsaturated soil is often reflected in the thermal hydraulic (THM) coupling model. Many THM models for unsaturated soil are generally established based on the thermal model of saturated soil and the hydraulic model of unsaturated soil. This aspect will be discussed in the following text.

1.2.3 Thermohydro (TH) Coupling of Soil

The coupling of temperature field and seepage field in soil, namely hot water coupling, mainly manifests in two aspects: (1) When the pore fluid in the soil moves, its seepage movement promotes the movement of thermal energy in the soil through convection, coupled with heat exchange with the solid phase, resulting in the redistribution of temperature field in the soil. (2) The changes in temperature field within the soil will also cause changes in the temperature of the pore fluid environment, leading to water migration and movement, thereby altering the thermal and physical properties of the soil. Due to changes in the internal structural characteristics of the soil, the flow characteristics of pore fluids are affected, resulting in changes in the flow field. Through mutual interaction and complementarity, a dynamic stable state is ultimately achieved, forming the so-called hot water coupling.

The problem of water heat coupling is more common in the study of frozen soil. During the freeze-thaw process, water in the unfrozen area will move toward the freezing front, accompanied by the generation of latent heat of phase change, which will change the temperature field of the soil, and the changes in temperature field will also cause changes in water content field, thus forming a coupling effect of mutual influence. Some scholars [44, 45] proposed a corresponding water heat coupling model to describe the changes in soil properties during freeze-thaw processes by studying the water heat migration process in frozen soil.

1. The influence of temperature on permeability coefficient

The influence of temperature on permeability coefficient is not only related to the dynamic viscosity coefficient of pore fluid, but also to the internal structure of soil, mineral composition, and ion concentration and type in the fluid. The measurement of permeability coefficient is generally obtained based on the consolidation compression curve, also known as indirect testing. However, there are few direct tests for measuring soil permeability at different temperatures, including hard wall permeability tests and flexible wall permeability tests. Burghingoli et al. [46] and Tohhata et al. [47] analyzed the advantages and disadvantages of using indirect testing methods for permeability, and compared them with direct methods. Delage et al. [48] studied the variation of permeability with temperature by conducting constant head measurements at different temperatures. Indirect analysis based on consolidation parameter estimation indicates that indirect methods for estimating permeability from consolidation tests should be carefully considered. More scholars have adopted the hard wall permeability test method for permeability research, such as Morin et al. [49], who conducted constant head permeability tests on marine soils under different temperature conditions using rigid consolidation rings. Finn [50] believed that temperature has a relatively small effect on the viscosity of pore water, while Habibagahi [51] found through experimental research that heating can reduce the viscosity of water, leading to an increase in permeability coefficient. This conclusion was later confirmed by Rujivipat [52]. The experimental study by Ye et al. [53] showed that there is a certain difference between the estimated value and the measured value based on the permeability coefficient theory of pore water viscosity changes. As the temperature increases, the difference between the two becomes more significant.

The permeability coefficient of soil medium is generally expressed by the following equation:

$$k = K\frac{\rho g}{\mu} = K\frac{g}{\eta} \tag{1.4}$$

where, μ is the fluid viscosity coefficient; k is the permeability coefficient; K is the permeability; ρ is the fluid density; η is the hydrodynamic viscosity coefficient; g is the acceleration due to gravity.

From formulas (1.4), it can be seen that the influence of temperature on the flow characteristics mainly comes from its effect on the dynamic viscosity coefficient and fluid density. Due to the increase in temperature, the kinetic energy of the fluid increases, resulting in a decrease in viscosity coefficient. In addition, the surface tension of pore water also decreases with temperature, making it easier for the fluid to penetrate between soil particles. The change in fluid density will cause a change in porosity, resulting in a change in saturation and thus a change in permeability. Romero et al. [37] experimentally studied the relationship between permeability coefficient and temperature under fixed porosity and moisture content conditions:

$$\left.\frac{k_w(e,w,T)}{k_w(e,w,T_r)}\right|_{e,w} = \frac{\rho_w(T)\eta_w(T_r)}{\rho_w(T_r)\eta_w(T)} \approx 1 + \beta_T(T - T_r) \tag{1.5}$$

where, k_w is the permeability coefficient; e is the porosity ratio; w is the moisture content; T_r is the reference temperature; β_T is the fitting coefficient.

Van Genuchten [31] proposed an expression for the relative permeability coefficient of unsaturated soil that is related to saturation:

$$k_r = \frac{K}{K_s} = s_e^{\frac{1}{2}}\left[\frac{\int_0^{s_e}\frac{dx}{\psi(x)}}{\int_0^{1}\frac{dx}{\psi(x)}}\right]^2 \tag{1.6}$$

where, k_r is the relative permeability coefficient; K_s represents the permeability coefficient of saturated soil; K is the permeability coefficient of unsaturated soil; ψ is the matric suction of soil; s_e is the effective moisture content. This formula can also be expressed as

$$s_e = \frac{\theta - \theta_r}{\theta_s - \theta_r} \tag{1.7}$$

where, θ is the volumetric water content; θ_s is the saturated moisture content; θ_r is the residual moisture content.

Van Genuchten [31] also provided the relationship between effective saturation s_e and matric suction ψ, known as the famous vG model:

$$s_e = \left[\frac{1}{(\alpha\psi)^n + 1}\right]^{\frac{n-1}{n}} \tag{1.8}$$

In the equation, α and n are fitting parameters that are related to the soil water characteristic curve (SWCC) of unsaturated soil.

Grant and Salehzadeh [54] proposed a model to describe the relationship between suction and temperature:

$$\frac{s(w,T)}{s(w,T_r)} = \left(\frac{a_1 + b_1 T}{a_1 + b_1 T_r}\right)^{b_1} \tag{1.9}$$

where, s is the matric suction force; w is the moisture content; T_r is the reference temperature; a_1 and b_1 are empirical coefficients.

Romero et al. [37] conducted permeability tests on unsaturated clay at different temperatures and found that the permeability coefficient of unsaturated soil increases with temperature, and the greater the saturation, the more significant this effect becomes. When fully saturated, the ratio of the permeability coefficients of 80 °C to 22 °C is about 1.3, and when the saturation is less than 75%, the ratio of the two approaches 1. Zhang et al. [55] also reached a similar conclusion by measuring the permeability of bentonite and sand mixtures at different temperatures. The results showed that the permeability coefficient of the soil sample slightly increased with increasing temperature, but the effect of temperature on the permeability coefficient was not significant when the saturation was low.

2. The Influence of Temperature on Soil Water Characteristic Curve (SWCC)

The soil water characteristic curve (SWCC) is a curve that describes the relationship between suction and saturation or volumetric moisture content in unsaturated soil, and can be obtained through experiments. In general, suction increases with decreasing saturation, and the shape of SWCC varies among different types of soil. When the suction force is the same, the water storage capacity of fine-grained soil is greater, and the corresponding saturation is also greater. That is to say, soils with larger particles have a greater rate of change in suction during the drying process, as shown in Fig. 1.4 [56]. In engineering applications, SWCC is generally represented in semi-logarithmic coordinates and is divided into three regions during the drying process: capillary zone, transition zone, and residual zone, as shown in Fig. 1.5 [57]. The matric suction in the capillary zone starts from 0 and reaches the inlet value, indicating the beginning of the drying process. The intake value can be taken as the intersection point of the horizontal line at saturation and the slope of the curve. As

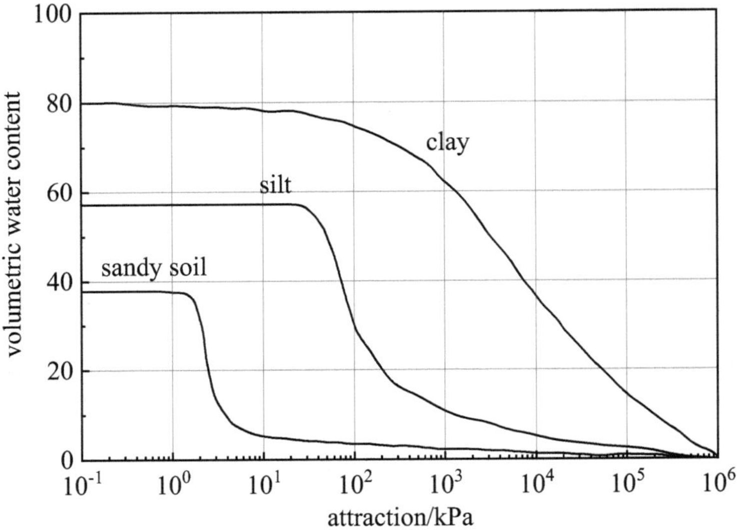

Fig. 1.4 Typical SWCCs for sand, silt, and clay [56]

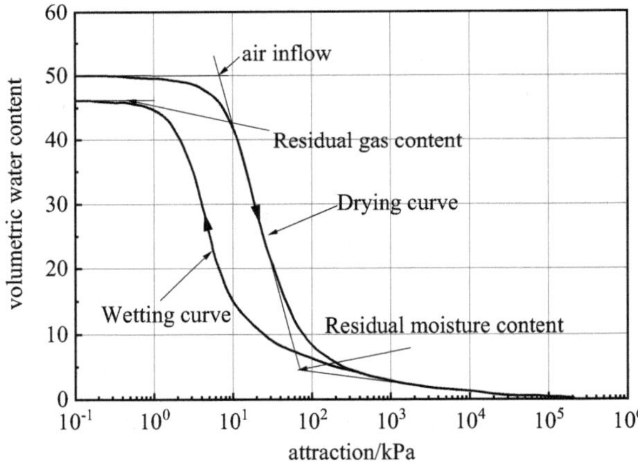

Fig. 1.5 Empirical procedure of determination of air-entry value and residual water content [57]

the drying process progresses, the suction force is already at a relatively high value when approaching the residual moisture content. If further drainage is required, a larger suction force is needed. Many scholars [54, 58–61] have derived the mathematical form of SWCC through fitting experimental data. In addition to the vG model (formula (1.8)) mentioned earlier, the model proposed by Fredlund and Xing [61] has also been widely recognized and applied, as shown below:

$$\theta = C(\psi)\theta_s \left[\frac{1}{\ln\left[e + (\psi / a)^n \right]} \right]^m \tag{1.10}$$

$$C(\psi) = \left[1 - \frac{\ln(1 + \psi / \psi_r)}{\ln\left(1 + 10^6 / \psi_r\right)} \right] \tag{1.11}$$

where, a, m, and n are the fitting parameters of the model; e is a natural logarithm constant; $C(\psi)$ is a correction factor that enables the model to characterize a suction value of 10^6kPa when the moisture content is 0; ψ_r is the suction value at the residual moisture content state.

The relationship between temperature and soil water characteristic curve is generally reflected by adding temperature influencing factors to the existing SWCC model. Grant and Salehzadeh [54] proposed a soil water characteristic curve model considering the influence of temperature by fitting the relationship between suction and temperature (Eq. (1.9)) on the basis of the vG model, which is expressed as follows:

$$s_r = \left[1 + \left(\frac{b_2 + T_0}{b_2 + T} \frac{s}{a} \right)^n \right]^{-m} \tag{1.12}$$

where, b_2, a, m, and n are fitting parameters, and $b_2 = -439.5$ °C is a constant; T_0 is the ambient temperature.

Hopmans and Dane [61] divided the pore water in soil into two types: isolated water and continuous water. When the temperature rises, the interfacial tension decreases, and capillaries cannot constrain all isolated water, causing some isolated water to combine with continuous water, thereby increasing the amount of continuous water. Wang et al. [62] classified pore water into free water, weakly bound water, and strongly bound water based on the internal structure of the soil. When heated, weakly bound water will convert to free water, causing a change in SWCC. Elkehsky [63] analyzed the effects of temperature, plasticity index, and other parameters on SWCC based on experimental data and found an inverse relationship between temperature and suction. The effect of temperature was more pronounced at lower saturation levels (<20%). Romero et al. [37] studied the effect of temperature on clay SWCC. The experiment was divided into two parts: low moisture soil samples were subjected to matric suction using steam equilibrium technique; when the soil sample has a high moisture content, axial translation technique is used to apply matric suction. Research has found that for soil samples with lower moisture content, the moisture content slightly decreases with increasing temperature. For soil samples with higher moisture content, the effect of temperature on SWCC is more pronounced at lower suction, and the moisture content also slightly decreases with increasing temperature.

SWCC has a significant impact on the permeability characteristics of unsaturated soil. Compared with the permeability coefficient, SWCC is easier to measure and is a key factor in simulating the permeability, volumetric strain, and shear characteristics of unsaturated soil [64, 65].

1.2.4 Thermo-Hydro-Mechanical (THM) Coupling of Soil

There have been many studies on the THM coupling problem in soil from three aspects: theoretical analysis, experiments, and numerical simulations. Early THM coupling problems were usually analyzed using porous media theory, including Biot isothermal consolidation theory, classical elastoplastic theory, and mixture theory. Many scholars have proposed THM coupling models for saturated or unsaturated soils based on these theories and different research backgrounds. Geraminegad and Saxena [66] proposed a THM coupled model that considers the deformation characteristics of soil, which can reflect the volume changes caused by changes in air pressure and suction. Noorishad and Tsang [67] first proposed the governing equation for THM coupling in saturated rock and soil based on the extended Biot consolidation theory. Gatmiri and Delage [68] proposed a THM coupled model that takes into account the nonlinear elastic deformation, compressibility, thermal expansion, and thermal convection of soil media. Khalili and Loret [69] proposed an elastoplastic model based on the Barcelona Basic Model (BBM) model to explain the effects of suction and temperature on the yield surface. Wu et al. [70] studied the effect of temperature on the hydraulic properties of unsaturated soils and found that the maximum yield curves of pre-consolidation pressure and suction would decrease with increasing temperature, i.e., thermal softening. They proposed a THM coupled model for unsaturated soils. Tong et al. [71] proposed a THM coupling model for unsaturated soils based on mixture theory, considering solid-phase deformation, liquid-phase and gas-phase transport, and heat conduction. Dumont et al. [72] proposed a simple THM coupling model for unsaturated soils using the concept of effective stress and fewer parameters. Gens and Olivella [73] developed a theoretical framework that can describe the THM coupling characteristics of soil, and this theory can be used to analyze some engineering THM coupling problems through numerical calculations. Cai et al. [74] proposed a fully coupled theoretical framework for unsaturated soil THM based on continuum mechanics and mixture theory.

The numerical calculation methods for THM coupling problems mainly include finite element method, finite difference method, discrete element method, and SPH calculation method. Wang et al. [75] used the parallel Newton Raphson algorithm with domain decomposition technique to solve the THM coupling problem in porous media with two-phase flow of water and gas. This algorithm considers subregions as composite units, and the equations of each subregion only consider the degrees of freedom of its boundary nodes. As long as the unknown variables of the boundary nodes of the subregion are determined, the internal variables can be independently solved. Rutqvist et al. [76] proposed the governing equations for the

THM coupling process in porous media and compared and analyzed the frameworks of four finite element calculation methods. They found that the main difference lies in the treatment of the gas phase. Li et al. [77] used a mixed finite element method to perform THM coupled numerical analysis on porous media elastoplastic materials. Lu et al. [78] used the finite difference method to simulate the THM coupling problem of clay barriers and studied the effect of radioactive waste heat release on soil deformation. Analysis suggests that due to the complex geometric and physical state of clay, it is very difficult to simulate it using finite difference method. It is recommended to combine it with finite element method for simulation calculations.

1.2.5 Current Research Status of Gassy Soil

Gas containing soil generally refers to soil with enclosed internal gas, which is the main difference from unsaturated soil. Different types of gas bearing soils have different forms of gas occurrence [79]: for coarse-grained soils (such as sand), the size of bubbles is relatively small compared to soil particles, and the bubbles are surrounded by pore water and free within the pores of the soil skeleton, as shown in Fig. 1.6a; In the gas storage sand layer rich in natural gas, the sealed gas is in a connected state, the saturation of the soil is low, and the internal pressure is higher than atmospheric pressure, as shown in Fig. 1.6b. For fine-grained soil (such as soft

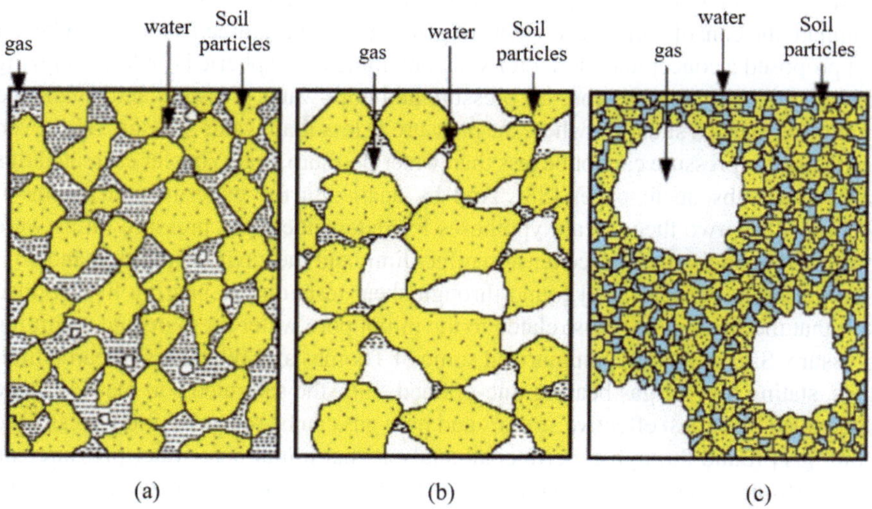

Fig. 1.6 Three forms of gassy soil: (**a**) In coarse-grained soil, air bubbles are located in the pore of soil skeleton, surrounded by liquid phase, and the saturation is high. (**b**) In coarse-grained soil, air is connected and closed, and the saturation is low. (**c**) In fine-grained soil, air bubbles are larger than soil particles, and air bubbles are surrounded by saturated soil matrix, and the saturation is high [79]

clay), the size of bubbles is relatively large compared to soil particles, and they exist in the form of discrete bubbles in a completely saturated soil matrix, as shown in Fig. 1.6c. The mechanical properties of different types of gas bearing soils are also different, such as the gas bearing sand soil shown in Fig. 1.6a. The presence of gas does not affect the deformation of the soil skeleton, and the effective stress principle of the Taisha foundation can still be used, but the presence of gas increases the compressibility of the pore fluid [80]. The presence of gas in the gas containing soft clay shown in Fig. 1.6c will affect the deformation of the soil skeleton, and the effective stress principle of Taisha foundation is no longer applicable. When analyzing its mechanical properties, the influence of gas should be considered [1]. Early studies [81–86] mostly focused on the engineering mechanical properties of gas bearing soils in marine areas, including the influence of gas volume content on shear strength, bearing capacity, and stability. For example, Christian and Cranston [84] found that the presence of gas can significantly alter the mechanical properties of soil, such as increasing the compressibility of pore fluids, changing the stiffness, shear strength, and permeability of soil, which can easily cause some engineering disasters in gas bearing formations.

1. Experimental Study

The research on gas bearing soil started earlier abroad. Thomas [87] conducted one-dimensional consolidation tests on fine-grained soil containing large bubbles and found that during the consolidation process, the internal gas pressure increased linearly with the increase of total loading stress. However, during the initial stage of unloading, the gas pressure remained constant and did not change with the decrease of stress. And based on this, a dual compressibility model was proposed, which can simulate the consolidation test results of gas bearing fine-grained soil well. Wheeler [1] proposed a conceptual model for soil containing atmospheric bubbles, analyzing the relationship between pore air pressure and pore water pressure, and concluded that there exists a situation where air pressure is less than water pressure. However, in reality, air pressure cannot be less than water pressure, and Wheeler corrected this error in a subsequent paper [88]. And by analyzing the experimental results of Thomas [87], two theoretical hypotheses for independent gas pressure limits were proposed, namely the surface tension effect limit and the cavity expansion and contraction limit. Wheeler [89] found through shear tests on gas bearing fine-grained soil that the shear strength is related to the initial pore water pressure and confining pressure. Sills et al. [90] summarized some of Thomas and Wheeler's research findings, stating that for gas bearing fine-grained soils, the saturated soil matrix can be controlled by "quasi effective stress", and pore gas can be related to total stress. Rad et al. [91] found through experimental analysis that initial pore water pressure can have a strong impact on soil, and the lower the pore pressure, the generally lower the undrained strength. Gardner [92] measured the compressive wave velocity in gas bearing soil, analyzed its attenuation law, and studied the structural characteristics that cause unique acoustic responses. Grozic [93] conducted a series of shear tests on soft gas bearing sand and analyzed the influence of gas on the mechanical properties of gas bearing sand. Wichman et al. [94] developed a calculation program for the self-weight consolidation of gas contaminated soil based on various theoretical assumptions, and verified these assumptions through small-scale column experiments simulating the generation of biogas. Sill and Gonzalez [95] also conducted experimental research on the self-weight consolidation behavior of gas containing soft soil. The self-weight stress did not exceed 1 kPa, and high stress was achieved through hydraulic gradient. At this stage, the influence of gas in the soil on soil strength decreased. Breul et al. [96] proposed a technique based on geological endoscopy and automatic image analysis to estimate the characteristics of the internal structure and gas content of gas bearing soil, and presented the results of laboratory and field experiments. Amaratunga and Grozie [97] studied the pore pressure response and volumetric strain of loose gas bearing sandy soil under undrained unloading stress path through experiments, and proved experimentally and theoretically that it exists in free and/or dissolved form in undrained unloading stress path.

2. Theoretical Research

In terms of theoretical research, Dusseault [98] derived the compressibility of pore fluids and coupled it with the compressibility of soil skeletons to provide a nonlinear elastic constitutive model that can describe the deformation characteristics of gas bearing sandy soils. Wheeler [1] studied the consolidation and shear characteristics of gas containing fine-grained soil through experiments and proposed a conceptual model of gas containing porous soil. Gas containing fine-grained soil is regarded as composed of a "gas cavity" and its surrounding saturated soil matrix, where the mechanical properties of the saturated matrix depend on the difference between total stress and pore water pressure, similar to the effective stress in general saturated soil. It is also pointed out that the mechanical properties of gas containing soil need to consider the influence of gas pressure and gas volume. This model was later developed by many scholars [90] due to its simplicity and ability to describe some mechanical properties of gas bearing fine-grained soils; Sultan et al. [99, 100] and Hong et al., [101] adopted it. Wheeler [89] proposed the large bubble model and provided expressions for the upper and lower boundaries of undrained shear strength and elastic modulus of gas bearing fine-grained soil. Mathiroban [102] studied the undrained shear characteristics of loose gas bearing sandy soil through experiments and proposed an elastic-plastic constitutive model to describe the changes in pore pressure and shear stress of gas bearing sandy soil during undrained shear process. Grozic et al. [93, 103, 104] conducted a series of experimental studies on gas bearing sand and conducted theoretical analysis based on this. A constitutive model that can describe the basic mechanical properties of gas bearing sand was proposed on the basis of a saturated sand model. Atigh et al. [105] proposed an effective stress method based on the elastic-plastic stress-strain relationship to simulate sand liquefaction flow, and evaluated the mechanical behavior of loose saturated and gas bearing sand using a fully coupled analysis method. Grozic [106] proposed a constitutive model that reflects the undrained shear strength of gas bearing fine-grained soil. The simulation results indicate that although the presence of gas has a favorable effect on the shear strength of gas bearing fine-grained soil, the peak strength is only reached under significant deformation. Sultan and Garziglia

[100] proposed a constitutive model that can describe the basic mechanical properties of gas bearing fine-grained soil by introducing gas failure parameters based on the Cambridge model and addressing the impact of gas bubble enlargement and aggregation on fine-grained soil.

1.3 Issues

From the above summary of existing research results, it can be seen that valuable research achievements have been made in the study of multi-field coupling in saturated soil, unsaturated soil, and gas bearing soil, especially in the research on the influence of temperature field on soil mechanical properties, including experimental and theoretical model studies, providing a solid foundation for further in-depth research. However, there are still some problems in current research, which are also new challenges faced by the development of multi-field coupling models for soil:

1. In the research of multi-field coupling models, the coupling characteristics are often reflected by adding coupling terms. However, in the corresponding constitutive equations, the coupling effects of other fields are rarely involved. The resulting coupling models are not truly multi-field coupling and do not fundamentally reflect the characteristics of multi-field coupling. In addition, some of the coupling coefficients involved in the model may not have specific physical meanings and are difficult to determine through experiments.
2. Most existing soil theoretical models are based on traditional constitutive theories, such as the widely used classical elastoplastic mechanics theory. However, the non-correlated flow model developed based on the theory of elastic-plastic mechanics to describe the constitutive characteristics of rock and soil does not conform to Drucker's postulate, and often requires various modifications to reflect the complex mechanical properties of rock and soil. There are also some coupled models based on thermodynamic theory, which define the free energy function and dissipation potential function, and use thermodynamic principles to obtain the flow rules and yield criteria of geotechnical materials, avoiding the theoretical defects of plastic potential function in elastic-plastic mechanics models, also known as superplastic models. However, this theoretical model does not deviate from the construction approach of classical elastic-plastic mechanics models, and still requires the definition of concepts such as yield criteria and flow rules. The yield criteria or hardening parameters involved are often fitted through some indoor test results, lacking a rigorous and scientific theoretical basis, and the resulting models may only have good simulation effects on a certain test result. Therefore, there is still a lack of a unified constitutive theory that is more universally applicable to geotechnical materials.
3. The research on soil mainly focuses on saturated and unsaturated soils, and there is a lack of relevant theoretical and experimental studies on a special type of gas bearing soil. Especially in terms of the influence of temperature on the mechanical

properties of gas bearing soils, there are almost no relevant literature reports. There is an urgent need to establish and develop research on the mechanical properties and multi-field coupling characteristics of gas bearing soils.

References

1. Wheeler SJ (1988a) A conceptual model for soils containing large gas bubbles. Géotechnique 38(3):389–397
2. Sobkowikz JC (1982) The mechanics of gassy sediments. University of Alberta, Alberta
3. Bishop AW, Alpan I, Blight GE, Donald IB (1960) Factors controlling the strength of partly saturated cohesive soils. In: Research conference on shear strength of cohesive soils. American Society of Civil Engineers, New York, pp 503–532
4. Ohnishi Y, Shibata H, Kobayashi A (1987) Development of finite element code for the analysis of coupled thermo-hydro-mechanical behaviors of a saturated-unsaturated medium. In: Coupled processes associated with nuclear waste repositories, pp 679–697
5. Vaziri HH (1996) Theory and application of a fully coupled thermo-hydro-mechanical finite element model. Comput Struct 61(1):131–146
6. Gawin D, Schrefler BA, Galindo M (1996) Thermo-hydro-mechanical analysis of partially saturated porous materials. Eng Comput 13(7):113–143
7. Sharma R (1998) Mechanical behavior of unsaturated highly expansive clay. University of Oxford, Oxford
8. Alonso EE, Gens A, Josa A (1990) A constitutive model for partially saturated soils. Géotechnique 40(3):405–430
9. Wheeler SJ, Sharma RS, Buisson MSR (2003) Coupling of hydraulic hysteresis and stress-strain behaviour in unsaturated soils. Geotech 53(1):41–54
10. Wheeler SJ, Sivakumar V (1995) An elasto-plastic critical state framework for unsaturated soil. Géotechnique 45(1):35–53
11. Cui YJ, Delage P (1996) Yielding and plastic behavior of an unsaturated compacted silt. Géotechnique 46(2):291–311
12. Huang S, Barbour SL, Fredlund DG (1998) Development and verification of a coefficient of permeability function for a deformable unsaturated soil. Can Geotech J 35(3):411–425
13. Tarantino A (2009) A water retention model for deformable soils. Géotechnique 59(9):751–762
14. Sheng D, Sloan S, Gens A (2004) A constitutive model for unsaturated soils: thermomechanical and computational aspects. Comput Mech 33:453–465
15. Sun D, Sheng D, Sloan S (2007) Elastoplastic modelling of hydraulic and stress-strain behaviour of unsaturated soils. Mech Mater 39:212–221
16. Li XS (2007) Thermodynamics-based constitutive framework for unsaturated soils_1: theory. Géotechnique 57(5):411–422
17. Khalili N, Habte M, Zargarbash S (2008) A fully coupled flow deformation model for cyclic analysis of unsaturated soils including hydraulic and mechanical hystereses. Comput Geotech 35(6):872–889
18. Wong K, Mašín D (2014) Coupled hydro-mechanical model for partially saturated soils predicting small strain stiffness. Comput Geotech 61:355–369
19. Georgiadis K, Potts DM, Zdravkovic L (2005) Three-dimensional constitutive model for partially and fully saturated soils. Int J Geomech 5(3):244–255
20. Wheeler SJ (1996) Inclusion of speciefic water volume within an elasto-plastic model for unsaturated soils. Can Geotech J 33(1):42–57
21. Sun D, Sun W, Xiang L (2010) Effect of degree of saturation on mechanical behavior of unsaturated soils and its elastoplastic simulation. Comput Geotech 37(5):678–688

22. Muraleetharan K, Liu C, Wei C, Kibbey T, Chen L (2009) An elastoplastic framework for coupling hydraulic and mechanical behavior of unsaturated soils. Int J Plast 25(3):473–490
23. Ghasemzadeh H, Amiri SAG (2013) A hydro-mechanical elastoplastic model for unsaturated soils under isotropic loading conditions. Comput Geotech 51:91–100
24. Liu C, Muraleetharan K (2012) Coupled hydro-mechanical elastoplastic constitutive model for unsaturated sands and silts_I: formulation. Int J Geomech 12(3):239–247
25. Liu C, Muraleetharan K (2012) Coupled hydro-mechanical elastoplastic constitutive model for unsaturated sands and silts_II: integration, calibration, and validation. Int J Geomech 12(3):248–259
26. Gray WG, Schrefler BA (2001) Thermodynamic approach to effective stress in partially saturated porous media. Eur J Mech 20(4):521–538
27. Coussy O, Pereira JM, Vaunat J (2010) Revisiting the thermodynamics of hardening plasticity for unsaturated soils. Comput Geotech 37(1):207–215
28. Zhao CG, Liu Y, Gao FP (2010) Work and energy equations and the principle of generalized effective stress for unsaturated soils. Int J Numer Anal Methods Geomech 34(9):920–936
29. Hu R, Chen YF, Liu HH, Zhou CB (2015) A coupled stress-strain and hydraulic hysteresis model for unsaturated soils: thermodynamic analysis and model evaluation. Comput Geotech 63:159–170
30. Gallipoli D (2012) A hysteretic soil-water retention model accounting for cyclic variations of suction and void ratio. Geotechnique 62(7):605–616
31. van Genuchten MT (1980) A closed-form equation for predicting the hydraulic conductivity of unsaturated soils. Soil Sci Soc Am J 44:892–898
32. Assouline S (2006) Modeling the relationship between soil bulk density and the water retention curve. Vadose Zone J 5(2):554–563
33. Sun DA, Sheng DC, Cui HB, Sloan SW (2007) A density-dependent elastoplastic hydro-mechanical model for unsaturated compacted soils. Int J Numer Anal Methods Geomech 31(11):1257–1279
34. Zhou AN, Sheng D, Carter JP (2015) Modelling the effect of initial density on soil-water characteristic curves. Geotechnique 62(8):669–680
35. Cui YJ, Sultan N, Delage P (2000) A thermomechanical model for saturated clays. Can Geotech J 37(3):607–620
36. Abuel-Naga H, Bergado D, Bouazza A, Ramana G (2007) Volume change behaviour of saturated clays under drained heating conditions: experimental results and constitutive modeling. Can Geotech J 44(8):942–956
37. Romero E, Gens A, Lloret A (2003) Suction effects on a compacted clay under non-isothermal conditions. Geotech 53(1):65–81
38. Abuel-Naga H, Bergado D, Bouazza A, Pender M (2009) Thermomechanical model for saturated clays. Géotechnique 59(3):273–278
39. Hueckel T, Baldi G (1990) Thermoplasticity of saturated clays: experimental constitutive study. J Geotech Eng 116(12):1778–1796
40. He LW, Jin ZH (2011) Effects of local thermal non-equilibrium on the pore pressure and thermal stresses around a spherical cavity in a porous medium. Int J Eng Sci 49:240–252
41. Hong P, Pereira J, Tang A, Cui YJ (2013) On some advanced thermo-mechanical models for saturated clays. Int J Numer Anal Methods Geomech 37(17):2952–2971
42. Laloui L, François B (2009) ACMEG-T: Soil Thermoplasticity Model. J Eng Mech 135(9):932–944
43. Zhou C, Ng CWW (2015) A thermomechanical model for saturated soil at small and large strains. Can Geotech J 52(8):1101–1110
44. Zhou J, Wei C, Li D, Wei H (2014) A moving-pump model for water migration in unsaturated freezing soil. Cold Reg Sci Technol 104–105:14–22
45. Li S, Zhan H, Lai Y, Sun Z, Pei W (2014) The coupled moisture-heat process of permafrost around a thermokarst pond in Qinghai-Tibet plateau under global warming. J Geophys Res Earth Surf 119(4):836–853

46. Burghignoli A, Desideri A, Miliziano S (1995) Discussion on volume change of clays induced by heating as observed in consolidation test. Soils Found 35(3):122–124
47. Towhata I, Kuntiwattanakul P, Seko I, Ohishi K (1995) Discussion on volume change of days induced by heating as observed in consolidation tests. Soils Found 35(3):124–127
48. Delage P, Sultan N, Cui YJ (2000) On the thermal consolidation of boom clay. Can Geotech J 37(4):343–354
49. Morin R, Silva AJ (1984) The effects of high pressure and high temperature on some physical properties of ocean sediments. J Geophys Res 89(Bl):511–526
50. Finn FN (1951) The effect of temperature on the consolidation of soils. In: American Society for Testing and Materials, Special Technical Publication, vol 126, pp 65–72
51. Habibagahi K (1977) Temperature effect and the concept of effective void ratio. Indian Geotech J 7:14–34
52. Abuel-Naga HM, Bergado DT, Soralump S, Rujivipat P (2005) Thermal consolidation of soft Bangkok clay. Lowland Technol Int 7(1):13–21
53. Ye WM, Wan M, Chen B, Chen YG, Cui YJ, Wang J (2013) Temperature effects on the swelling pressure and saturated hydraulic conductivity of the compacted GMZ01 bentonite. Environ Earth Sci 68(1):281–288
54. Grant S, Salehzadeh A (1996) Calculation of temperature effects on wetting coefficients of porous solids and their capillary pressure functions. Water Resour Res 32(2):261–270
55. Zhang M, Zhang H, Zhou L, Jia L (2013) Temperature effects on unsaturated hydraulic property of bentonite-sand buffer backfilling mixtures. J Wuhan Univ Technol-Mater Sci Ed 28(3):487–493
56. Vanapalli SK, Sillers WS, Fredlund MD (1998) The meaning and relevance of residual state to unsaturated soils. In: Proceedings of the fifty-first Canadian geotechnical conference, pp 1–8
57. Lu YT (2015) Temperature effect on unsaturated hydraulic properties of two fine-grained soils and its influence on moisture movement under an airfield test facility. Arizona State University, Phoenix
58. Campbell GS (1974) A simple method for determining unsaturated conductivity from moisture retention data. Soil Sci J 177(6):311–314
59. McKee CR, Bumb AC (1987) Flow-testing coalbed methane production wells in the presence of water and gas. Soc Pet Eng Form Eval 2(4):599–608
60. Fredlund DG, Xing A (1994) Equations for the soil-water characteristic curve. Can Geotech J 31(3):521–532
61. Hopmans JW, Dane JH (1986) Temperature dependence of soil hydraulic properties. Soil Sci Soc Am J 50(1):4–9
62. Wang X, Shao H, Hesser J, Zhang C, Wang W (2014) Numerical analysis of thermal impact on hydro-mechanical properties of clay. J Rock Mech Geotech Eng 6(5):405–416
63. Elkehsky MM (2011) Temperature effect on the soil water retention characteristic. Arizona State University, Phoenix
64. Vanapalli SK, Fredlund DG, Pufahl DE, Clifton AW (1996) Model for the prediction of shear strength with respect to soil suction. Can Geotech J 33(3):379–392
65. Mualem Y (1986) Hydraulic conductivity of unsaturated soils: predictions and formulas. In: Methods of soil analysis, Part I: Physical and mineralogical methods, pp 799–823
66. Geraminegad M, Saxena SK (1986) A coupled thermoelastic model for saturated–unsaturated porous media. Géotechnique 36(4): 539–550
67. Noorishad J, Tsang CF (1996) ROCMAS simulator; a thermohydromechanical computer code. Dev Geotech Eng 79:551–558
68. Gatmiri B, Delage P (1997) A new formulation of fully coupled thermal-hydro-mechanical behavior of saturated porous media: numerical approach. Int J Numer Anal Methods Geomech 21(3):199–225
69. Khalili N, Loret B (2001) An elasto-plastic model for non-isothermal analysis of flow and deformation in unsaturated porous media: formulation. Int J Solids Struct 38(46–47): 8305–8330

70. Wu W, Li X, Charlier R, Collin F (2004) A thermo-hydro-mechanical constitutive model and its numerical modelling for unsaturated soils. Comput Geotech 31:155–167
71. Tong FG, Jing LR, Zimmerman RW (2010) A fully coupled thermo-hydro-mechanical model for simulating multiphase flow, deformation and heat transfer in buffer material and rock masses. Int J Rock Mech Min Sci 47:205–217
72. Dumont M, Taibi S, Fleureau JM, Bekr NA, Saouab A (2010) Modelling the effect of temperature on unsaturated soil behavior. Compt Rendus Geosci 342(12):892–900
73. Gens A, Olivella S (2001) THM phenomena in saturated and unsaturated porous media: fundamentals and formulation. Rev Fr Genie Civil 5(6):693–717
74. Cai GQ, Zhao CG, Sheng DC, Zhou AN (2014) Formulation of thermo-hydro-mechanical coupling behavior of unsaturated soils based on hybrid mixture theory. Acta Mech Sin 30(4):559–568
75. Wang X, Gawin D, Schrefler BA (1996) A parallel algorithm for thermo-hydro-mechanical analysis of deforming porous media. Comput Mech 19(1):94–104
76. Rutqvist J, Börgesson L, Chijimatsu M, Kobayashi A, Jing L, Nguyen TS, Noorishad J, Tsang CF (2001) Thermohydromechanics of partially saturated geological media: governing equations and formulation of four finite element models. Int J Rock Mech Min Sci 38(1):105–127
77. Li XK, Liu Z, Lewis RW (2005) Mixed finite element method for coupled thermo-hydro-mechanical process in poro-elasto-plastic media at large strains. Int J Numer Methods Eng 64(5):667–708
78. Lu Y, Wu X, Cui Y (2010) On the thermo-mechanical properties of unsaturated soils. J Rock Mech Geotech Eng 2(2):143–148
79. Wang Y, Kong LW, Wang YL, Wang M, Cai KJ (2018) Deformation analysis of shallow gas-bearing ground from controlled gas release in Hangzhou Bay of China. Int J Geomech 18(1):04017122
80. Finno RJ, Zhang Y, Buscarnera G (2017) Experimental validation of Terzaghi's effective stress principle for gassy sand. J Geotech Geoenviron Eng 143(12):04017092
81. Whelan T, Coleman JM, Suhayda JN, Roberts HH (1977) Acoustical penetration and shear strength in gas-charged sediment. Mar Geotechnol 2:147–159
82. Wheeler SJ (1991) An alternative framework for unsaturated soil behavior. Géotechnique 41(2):257–261
83. Chillarige AV (1995) Liquefaction and seabed instability in the Fraser River Delta. University of Alberta, Alberta
84. Christian HA, Cranston RE (1997) A methodology for detecting free gas in marine sediments. Can Geotech J 34(2):293–304
85. Brandes HG (1999) Predicted and measured geotechnical properties of gas-charged sediments. Int J Offshore Polar Eng 9(3):219–225
86. Pralle N, Külzer M, Gudehus G (2003) Experimental evidence on the role of gas in sediment liquefaction and mud volcanism. Geol Soc Lond Spec Publ 216(1):159–171
87. Thomas SD (1987) The consolidation behavior of gassy soil. Oxford University, Oxford
88. Wheeler SJ, Sham WK, Thomas SD (1990) Gas pressure in unsaturated offshore soil. Can Geotech J 27(1):79–89
89. Wheeler SJ (1988b) The undrained shear strength of soils containing large gas bubbles. Géotechnique 38(3):399–413
90. Sills GC, Wheeler SJ, Thomas SD, Gardner TN (1991) Behavior of offshore soil containing gas-bubbles. Géotechnique 41(2):227–241
91. Rad NS, Vianna AJD, Berre T (1994) Gas in soils. II: effect of gas on undrained static and cyclic strength of sand. J Geotech Eng 120(4):716–737
92. Gardner TN (1999) An acoustic stud of soils that model seabed sediments containing gas bubbles. J Acoust Soc Am 107(1):163–176
93. Grozic JLH (1999) The behavior of loose gassy sand and its susceptibility to liquefaction. University of Alberta, Alberta

94. Wiehman BG, Sills GC, Gonzalez R (2000) Experimental validation of a finite strain theory for gassy mud. Can Geotech J 37(6):1227–1240
95. Sills GC, Gonzalez R (2001) Consolidation of naturally gassy soft soil. Géotechnique 51(7):629–639
96. Breul P, Haddani Y, Gourves R (2008) On site characterization and air content evaluation of coastal soils by image analysis to estimate liquefaction risk. Can Geotech J 45(12):1723–1732
97. Amaratunga A, Grozie JL (2009) On the undrained unloading behaviour of gassy sands. Can Geotech J 46(11):1267–1276
98. Duessault MB (1979) Undrained volume and stress change behaviour of unsaturated very dense sands. Can Geotech J 16(4):627–640
99. Sultan N, De Gennaro V, Puech A (2012) Mechanical behavior of gas-charged marine plastic sediments. Géotechnique 62(9):751–766
100. Sultan N, Garziglia S (2014) Mechanical behavior of gas-charged fine sediments: model formulation and calibration. Geotechnique 64(11):851–864
101. Hong Y, Wang LZ, Charles WW, Ng, Yang B (2017) Effect of initial pore pressure on undrained shear behavior of fine-grained gassy soil. Can Geotech J 54(11):1592–1600
102. Mathiroban S (2004) The undrained response of loose gassy sand. University of Calgary, Calgary
103. Grozic JLH, Robertson PK, Morgestern NR (1999) The behavior of loose gassy sand. Can Geotech J 36(3):482–492
104. Grozic JLH, Robertson PK, Morgenstem NR (2000) Cyclic liquefaction of loose gassy sand. Can Geotech J 37(4):843–856
105. Atigh E, Byrne PM (2004) Liquefaction flow of submarine slopes under partially undrained conditions: an effective stress approach. Can Geotech J 41(l):154–165
106. Grozic JLH, Nadim F, Kvalstad TJ (2005) On the undrained shear strength of gassy clays. Comput Geotech 32(7):483–490

Chapter 2
The Multi-Field Coupled Constitutive Model of Unsaturated Soil Based on Granular Thermodynamics Theory

2.1 Introduction

The development of the unsaturated soil thermo-hydro-mechanical (THM) coupling model plays a very important role in solving certain geotechnical engineering problems and has been a topic of significant research in recent years [1–8]. Many classic elastoplastic models for unsaturated soils have been developed within the framework of critical state theory [9–14]. For instance, Alonso et al. [10] proposed the famous Barcelona Basic Model (BBM) model by using the matrix suction and net stress as dual stress variables. The BBM model describes the stress-strain relationship with the modified Cambridge model and captures the yield behavior of unsaturated soils using the wetting-collapse yield curve (LC yield). As a classic model, it has been widely cited and improved by researchers [15, 16]. Sheng et al. [17] proposed a model (SFG) to describe the volumetric changes in unsaturated soils, which can predict deformations during drying and loading processes. However, many of these models use dual stress state variables, considering the effect of suction but neglecting the impact of saturation on the strength and deformation of unsaturated soils. A large amount of experimental data shows that even when matrix suction and net stress are identical, the mechanical properties of soils with different saturation levels can still vary. Therefore, some scholars [18–20] have introduced saturation degree into constitutive models to describe its influence on soil properties. Additionally, the soil-water characteristic curve (SWCC) of unsaturated soils exhibits significant hysteresis, meaning that the curves for drying and wetting processes are different. Some researchers [21–24] have incorporated the SWCC into constitutive models, leading to the development of hydro-mechanical coupling models for unsaturated soils.

In order to simulate the effects of non-isothermal processes on the mechanical properties of unsaturated soils, some scholars [25–27] have proposed thermo-hydro-mechanical coupling models based on the classical models, reflecting the

G. Yang, *Multi-field Coupled Constitutive Model for Soils Based on Granular Thermodynamics Theory*, https://doi.org/10.1007/978-981-95-0249-3_2

thermo-hydro coupling effects by considering the influence of temperature on the plastic potential or yield surface. Some THM coupling models have been derived by extending thermal-mechanical models [28, 29].

The effect of suction on the hydraulic properties of unsaturated soils cannot be ignored. The coupling between soil-water characteristics and elastoplastic deformation is the core of unsaturated soil hydro-mechanical coupling constitutive models. Therefore, research into THM coupling in unsaturated soils inevitably involves studying the influence of temperature on the SWCC. Generally, models are based on common SWCC models (such as the van Genuchten [31] model), with temperature effects incorporated either through the impact on suction or by adding temperature factors to reflect the influence of temperature on the SWCC curve [32, 33]. In recent years, several scholars have studied THM coupling models for unsaturated soils from different perspectives. Bellia et al. [6], based on the concept of plastic boundaries and effective stress, proposed a THM coupling model. Unlike traditional models, this model uses effective saturation instead of suction and temperature to predict the mechanical behavior of unsaturated soils. Sánchez et al. [34] proposed a THM fully coupled model, based on a dual-porosity concept and incorporating existing mechanical dual-structure models, which considers the coupling of non-isothermal multiphase flow and mechanical properties. This model can simulate the thermo-hydro-mechanical coupling characteristics of expansive unsaturated clays. Abed [7] derived the theoretical formulation of fully coupled equilibrium control equations and extended physical mechanisms, proposing a THM coupling model framework to simulate some thermo-hydro-mechanical coupling features of unsaturated soils.

Many existing THM coupling models are based on classical elastoplastic theory, which is the most widely used constitutive model theory today, such as the famous Cambridge model. However, most of the constitutive equations used in these models are empirical formulas derived by fitting experimental results, lacking a strict theoretical foundation. As a result, the models may only be applicable to a specific type of soil under particular conditions and fail to effectively describe the coupling phenomena of soil under multiple field effects. To overcome the theoretical limitations of the classical elastoplastic models, some scholars [35–38] have proposed constitutive models based on thermodynamic principles. By analyzing the dissipative energy in the deformation process of the unsaturated soil skeleton and the capillary hysteresis process, and considering the thermodynamic potentials of solid, liquid, and gas phases, they defined the free energy function and dissipative potential function, establishing corresponding elastoplastic or superplastic constitutive models. These models provide a clearer physical meaning for concepts such as yield criteria and flow laws, and represent an important supplement to the classical elastoplastic-based models. However, they still follow the classical elastoplastic modeling approach, defining flow laws, loading-unloading criteria, yield functions, and other concepts.

Modeling methods based on fluid dynamics theory, which are grounded in classical non-equilibrium thermodynamics and physical conservation laws, have successfully been applied in materials such as polymers and crystals [39, 40]. Jiang and Liu [41] introduced the concept of granular entropy and granular temperature to obtain a theoretical constitutive model for single-phase particulate material. Particulate materials generally refer to aggregates of particles larger than 1 μm in size, which, besides having both microscopic and macroscopic spatial layers, also have a mesoscopic layer characterized by the distance between adjacent particles. This characteristic length is often related to particle size and can be determined through ultrasonic propagation methods [42]. The model considers both the macroscopic and microscopic behavior of particulate materials and is a multi-scale theoretical model. The assumptions made about macroscopic entropy, viscosity, and other physical quantities at the particle level have analogs at the particle level, tracing back to the particle gas dynamics theory established by Haff [43]. Jiang and Liu [44] compared this modeling theory with the superplastic modeling theory of Houlsby [45] outlining the advantages and disadvantages of the granular solid hydrodynamics (GSH) theory. Zhang et al. [46–48] extended the GSH theory to saturated soils, developing a THM coupling model for saturated clays that can simulate some thermo-hydro-mechanical coupling features. Jiang and Liu [49], based on fluid dynamics theory for particulate materials, proposed a simple free energy model for unsaturated soils and conducted comparative analysis of the constitutive modeling, free energy, and transfer coefficient modeling.

Although some constitutive relationships serve as important phenomenological empirical laws for describing certain experimental phenomena, their theoretical models do not belong to the framework of classical physics theory. Therefore, using them as the basis for continuous mechanics theory for geotechnical materials requires further scrutiny [49]. As geotechnical engineering problems become more complex, the factors required to derive constitutive relationships increase, making the establishment of constitutive models more difficult, and it is not guaranteed that the models will be effective. In contrast, fluid dynamics theory, based on classical physical theory, provides a very effective method for simulating certain behaviors of soils. From this perspective, attempting to use classical physics to solve the current dilemmas faced by unsaturated soils should be a valuable research approach. This chapter extends the GSH modeling method to the unsaturated soil three-phase mixture system based on the fluid dynamics theory of particulate materials and mixture theory, considering temperature and saturation changes that lead to corresponding energy dissipation processes at the particle level. Combined with an improved SWCC model that accounts for deformation and temperature effects, a multi-field coupled constitutive model for unsaturated soils is proposed. The model introduces the concepts of particle entropy and particle temperature, theoretically determining the dissipative structural composition and transfer coefficient relationships of the thermodynamic system. By combining the thermodynamic total differential equation and conservation equations of unsaturated soils and the entropy-increase equation, the thermodynamic identity is established. Through the thermodynamic identity and dissipation relations, the constitutive relationship for inelastic deformation is derived. Based on the transfer coefficient and energy function model, the dissipative mechanisms at the particle level of the soil can be related to macroscopic physical mechanical behaviors.

2.2 Granular Solid Hydrodynamics Theory

2.2.1 Granular Entropy and Granular Temperature

From the perspective of material structure, the particulate matter of geotechnical materials, such as soil and rocks, significantly differs from common fluids like water and air, or typical solids like crystals and metals.

In addition to having both macroscopic and microscopic spatial scales, geotechnical materials have an additional particulate scale, also known as the mesoscopic scale. The interactions at this scale involve contact forces between particles. In other words, apart from the thermal motion of molecules at the microscopic level, there are also irregular movements at the particulate level, such as sliding, collision, and rolling of particles, which are referred to as the fluctuation motion between particles. In thermodynamics, the irregular, disordered motion of molecules at the microscopic level is expressed at the macroscopic level as thermodynamic entropy S or temperature T. Condensed matter physicists have drawn an analogy between the fluctuation motion between particles and the irregular, disordered motion of microscopic molecules, introducing the concept of particulate entropy S_g or particulate temperature T_g to describe the fluctuation motion at the particulate level. Particulate temperature $\mathrm{T_g}$ and particulate entropy S_g are a pair of thermodynamic conjugates, analogous to the macroscopic temperature T and entropy S. Particulate temperature T_g differs from the general concept of temperature T. For instance, in an isothermal condition (with temperature T constant), a stationary granular material unit would have no dissipation at the particulate level, so $T_g = 0$. Once disturbed by an external force, the particles will undergo collision, sliding, and rolling, leading to energy dissipation and inelastic deformation. As a result, T_g will change until a new equilibrium is reached, at which point T_g will return to zero. Throughout this period, the temperature T is assumed to remain constant.

Unlike the microscopic molecular thermal motion, the interactions between particles are generally inelastic, which leads to energy dissipation. Without continuous excitation, the fluctuation motion of the particles will decay as macroscopic energy is dissipated, until the fluctuations disappear and equilibrium is re-established. In the case of ordinary gases, once the molecular motion reaches the Maxwell velocity distribution, the gas is in thermodynamic equilibrium. However, in a particulate matter system, even if the particles reach the Maxwell velocity distribution, the kinetic energy carried by the particles will decay into the microscopic molecular motion during their interactions, accompanied by entropy generation. Therefore, the fluctuation motion at the mesoscopic scale cannot be directly attributed to the thermodynamic entropy S, and particulate entropy Sg needs to be introduced to reflect its existence and evolution. As shown in Fig. 2.1, the dissipative process will generate particulate entropy Sg or directly generate heat entropy S, since the collisions between particles are inelastic and will accompany entropy production. Eventually, particulate entropy Sg will transform into heat entropy S. For geotechnical materials with multiple scales, this approach is simple and effective, as it

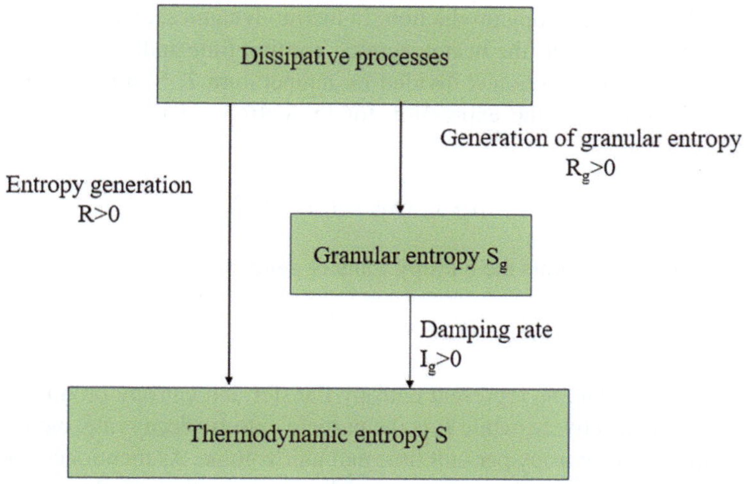

Fig. 2.1 Two-stage irreversible dissipation processes of granular system

captures the essential characteristics of the material's multi-scale behavior from the perspective of energy dissipation.

2.2.2 GSH Theory Framework

1. Conservation Equations and Entropy-Increase Equation

For a unit volume, a moving object possesses definitive values of mass, momentum, and energy at any given time. Conservation is a fundamental property of mass, momentum, and energy, and their conservation can be quantitatively described through differential equations.

$$\partial_t \rho + \nabla_k (\rho v_k) = 0 \tag{2.1}$$

$$\partial_t m_i + \nabla_j \sigma_{ij} = \rho g_i \tag{2.2}$$

$$\partial_t w + \nabla_k Q_k = \rho g_k v_k \tag{2.3}$$

where, ∂_t represents the partial-derivative with respect to time, ∇_k represents the partial-derivative with respect to space, such as $\nabla_k v_k = \frac{\partial v_x}{\partial x} + \frac{\partial v_y}{\partial y} + \frac{\partial v_z}{\partial z}$. ρ is the mass density, v_k is the velocity, m_i is the momentum density, σ_{ij} is the stress-tensor, g_i is the gravitational acceleration vector, w is the energy density, and Q_k is the energy flux, simply referred to as the energy flow.

In addition to mass, momentum, and energy, entropy is also a fundamental physical quantity possessed by all materials. Unlike conserved quantities, entropy can be generated, known as entropy production. In thermodynamics, entropy production is commonly represented as the heat generated per unit time and unit volume, which is the rate of heat production R divided by temperature T. Similar to the form of conservation equations, the expression for the entropy-increase equation can be derived.

$$\partial_t S + \nabla_k \left(S v_k - f_k \right) = R / T \tag{2.4}$$

In this case f_k represents the entropy transfer generated through heat conduction.

$$\partial_t S_g + \nabla_k \left(S_g v_k - f_k^g \right) = R_g / T_g - I_g \tag{2.5}$$

In this case, f_k^g and R_g represent entropy transfer and entropy production at the particle level, respectively, while I_g is the particle entropy decay rate, indicating the decrease in particle entropy per unit time and unit volume. As mentioned above, the kinetic energy carried by the moving particle system will decay into microscopic molecular motion, accompanied by entropy production. Therefore, the I_g-term must be subtracted from the particle entropy-increase equation.

The forms of the above conservation equations and the entropy-increase equation are independent of the specific material and do not contain any material parameters; they are general equations.

2. Thermodynamic equation

The set of thermodynamic independent variables for particulate matter is denoted as $\left\{ \rho, m_i, S, S_g, \varepsilon_{ij}^s \right\}$, and its corresponding conjugate variable is $\{\mu_c, v_i, T, T_g, \pi_{ij}\}$. Therefore, the total differential form of its thermodynamic energy can be obtained.

$$dw = \pi_{ij} d\varepsilon_{ij}^s + T_g dS_g + TdS + \mu_c d\rho + v_i dm_i \tag{2.6}$$

Here, w represents the total energy density; ε_{ij}^s represents the elastic strain; $m_i = \rho v_i$ represents momentum; v_i represents velocity; μ_c represents the chemical potential; and π_{ij} is generally referred to as the elastic stress.

According to the principle of maximum entropy at equilibrium, the dissipative force system is determined as $\{\pi_{ij}, T_g, T, \psi_{ij}\}$, and its corresponding dissipative flux is also $\left\{ Y_{ij}, I_g, q_k / T, \sigma_{ij}^v \right\}$. Here, q_k represents the heat conduction in the entropy generation equation; σ_{ij}^v represents the viscous stress; Y_{ij} represents the dissipative flux corresponding to the effective stress π_{ij}; and ψ_{ij} represents the deformation rate, $\psi_{ij} = -(\nabla_i v_j + \nabla_j v_i)/2$. Therefore, based on the Onsager [50] relation, the rate of entropy production (R) can be expressed as the sum of the products of dissipative forces and dissipative fluxes.

$$R = Y_{ij}\pi_{ij} + I_g T_g + \frac{q_k}{T}\nabla_i T + \sigma_{ij}^v \psi_{ij} \tag{2.7}$$

Similarly, if only the viscosity at the particle level is considered, the particle entropy production rate can be expressed as:

$$R_g = \sigma_{ij}^{vg} \psi_{ij} \tag{2.8}$$

In the nonequilibrium state region, which is not far from the equilibrium state, the dissipative flux can be expressed as a linear function of the dissipative force [51], as shown below:

$$\begin{bmatrix} Y_{ij} \\ I_g \\ q_i \\ \sigma_{ij}^{v} \end{bmatrix} = \begin{bmatrix} \eta_{ijkl} & 0 & 0 & 0 \\ 0 & \gamma & 0 & 0 \\ 0 & 0 & \kappa_{ik} & 0 \\ 0 & 0 & 0 & \lambda_{ijkl} \end{bmatrix} \begin{bmatrix} \pi_{kl} \\ T_g \\ \nabla_k T \\ \psi_{kl} \end{bmatrix} \tag{2.9}$$

The matrix in the middle is called the transfer coefficient matrix, and the coefficients within it represent the correlation coefficients between the corresponding dissipative flux and dissipative force. It is required to be a positive definite matrix, which not only satisfies Onsager's reciprocity [50], but also needs to comply with the Curie principle [51]. This means that the dissipative forces cannot have more symmetrical elements than the dissipative fluxes with which they are coupled. In other words, there is no coupling between fluxes and forces with different symmetry properties. This helps reduce the number of transfer coefficients. If two processes with significant interactions are present, off-diagonal coefficients can be added to the matrix, which are the correlation or coupling coefficients.

Similarly, the dissipative relation at the particle level is:

$$\sigma_{ij}^{vg} = \lambda_{ijkl}^{g} \psi_{kl} \tag{2.10}$$

Considering the isotropic case, η_{ijkl} and λ_{ijkl}^{g} can be expressed as:

$$\eta_{ijkl} = \eta_s T_g \delta_{il} \delta_{jk} + \left(\eta_v - \frac{\eta_s}{3} \right) T_g \delta_{ij} \delta_{lk} \tag{2.11}$$

$$\lambda_{ijkl}^{g} = \lambda_{sg} T_g \delta_{il} \delta_{jk} + \left(\lambda_{vg} - \frac{\lambda_{sg}}{3} \right) T_g \delta_{ij} \delta_{lk} \tag{2.12}$$

By combining the three fundamental conservation equations of mass, momentum, and energy, the entropy production equation, and Eqs. (2.6), (2.7), and (2.8), a thermodynamic identity can be constructed, from which the relationship can be obtained:

$$Y_{ij} = d_t \varepsilon_{ij}^{p} \tag{2.13}$$

where $d_t\varepsilon_{ij}^p = \psi_{ij} - d_t\varepsilon_{ij}^e$, the dissipative force corresponding to π_{ij}, is the plastic strain rate.

Based on the dissipative relation (2.9), the following can be obtained:

$$d_t\varepsilon_{ij}^p = \eta_{ijkl}\pi_{kl} = \eta_v T_g \varepsilon_{ij}^e \delta_{ij} + \eta_s T_g e_{ij}^e \tag{2.14}$$

If the elastic potential energy density function w_e is given, the expression for π_{ij} can be obtained:

$$\pi_{ij} = \partial w_e / \partial \varepsilon_{ij}^e \tag{2.15}$$

For example, the expression for the elastic potential energy density function w_e of particulate materials given by Jiang and Liu [41] is:

$$w_e = B\left(\varepsilon_v^e\right)^{0.5}\left[\frac{2}{5}\left(\varepsilon_v^e\right)^2 + \frac{1}{\xi}\left(\varepsilon_s^e\right)^2\right] \tag{2.16}$$

where B and ξ are material parameters, B describes the hardness or softness of the material and has the same dimension as stress; ξ is a material parameter similar to Poisson's ratio, related to the Coulomb yield angle. For dry sand, it can be taken as 5/3; 2/5 is a factor introduced for convenience; $\varepsilon_v^e = \varepsilon_{kk}^e$ is the elastic body strain; $\varepsilon_s^e = \sqrt{e_{ij}^e e_{ij}^e}$ is the elastic shear strain, $e_{ij}^e = \varepsilon_{ij}^e - \frac{1}{3}\varepsilon_{kk}^e\delta_{ij}$.

In addition, in the GSH theory, $w_g = b\rho^s T_g^2 / 2$ is the energy related to the fluctuating motion of soil particles [41], and b is a material parameter. By combining the thermodynamic energy total differential expression for particulate matter (2.6), the following can be obtained:

$$\rho\vartheta_g = \frac{\partial w_g}{\partial T_g} = \rho b T_g \tag{2.17}$$

where $\vartheta_g = S_g/\rho$ is the particle entropy capacity.

Thus, the relationship $\vartheta_g = bT_g$ can be obtained. By combining Eqs. (2.5), (2.8), (2.9), (2.10), and (2.14), the equation of motion for particle temperature can be derived as:

$$d_t T_g = \frac{\lambda_{sg} d_t e_{ij} d_t e_{ij}}{b\rho^s} + \frac{\lambda_{vg} d_t \varepsilon_v d_t \varepsilon_v}{b\rho^s} - \frac{\gamma T_g}{b\rho^s} \tag{2.18}$$

The above is a basic introduction to the GSH theory of single particle matter (without void fillers such as water or air). Although it is relatively complex, this comprehensive theoretical framework can be used to calculate the mechanical properties of particulate matter under any initial or boundary conditions [52, 53].

2.3 The Theoretical Framework of the Multi-Field Coupling Constitutive Model for Unsaturated Soil

2.3.1 Basic Assumptions

1. Unsaturated soil is a mixture composed of solid components (soil particles and their surrounding bound water, with bound water classified as a solid-phase component) S, liquid components (water, aqueous solutions, or other liquids in the pores) L, and gas components (air or mixed gases in the pores) G.
2. There is mass transfer between the liquid and gas phases of the unsaturated soil mixture. Only during heating, there is unidirectional mass transfer between the solid and liquid components (solid phase → liquid phase, i.e., bound water converting into free water), and the solid, liquid, and gas phases have the same temperature.
3. Both the solid and liquid components are incompressible, and their intrinsic density is constant. The gas components are compressible, and their intrinsic density can vary.

2.3.2 Determination of Thermodynamic State Variables

The physical state of a unit cell Newtonian fluid can, according to thermodynamics, be characterized by its mass ρ, momentum $m_i = \rho v_i$, and entropy S. This implies that all other macroscopic physical quantities, such as energy w, velocity v_i, temperature T, chemical potential μ_c, etc., can be expressed as functions of these variables. In other words, $\{\rho, m_i, S\}$ forms a complete set of independent state variables.

For elastic materials, such as crystals, rocks, and other materials with broken spatial translation symmetry [54], an additional state variable related to deformation should be introduced, namely elastic strain ε_{ij}. It is important to note that the total strain ε_{ij}^e is generally not used directly as this state variable, as it may contain plastic strain and cannot form a one-to-one correspondence with the thermodynamic state [49]. For particulate materials, there is an additional mesoscopic level, requiring the introduction of a pseudo-entropy variable to reflect the motion at the mesoscopic level (relative sliding and rolling between particles) and its manifestation at the macroscopic level, i.e., particle entropy S_g. The generation of particle entropy is primarily induced by external stimuli such as load or temperature changes, and can be quantitatively described using a form similar to the entropy-increase equation. This is more intuitive and clear compared to the internal variables or internal time variables used in superplasticity models. Therefore, for particulate materials, the thermodynamic state variable is $\left\{\rho, m_i, S, \varepsilon_{ij}^e, S_g\right\}$.

Unsaturated soil is a three-phase mixture consisting of solid, liquid, and gas phases. Based on mixture theory, the thermodynamic state variables are taken as the densities of the three phases ρ^a, momentum $m_i^a = \rho^a v_i^a$ $\left(a = [S,L,G]\right)$, total entropy

$S = n^s S_s + n^L S_L + n^G S_G$, particle entropy S_g, and elastic strain ε_{ij}^e. Therefore, the thermodynamic state variable set for unsaturated soil can be represented as $\left\{\rho^a, m_i^a, S, S_g, \varepsilon_{ij}^e\right\}$, and its conjugate quantities are $\left\{\mu_{ca}, v_i^a, T, T_g, \pi_{ij}\right\}$, where μ_{ca} is the chemical potential of phase a, and π_{ij} is the effective stress of the unsaturated soil [55].

2.3.3 Fundamental Equations and Thermodynamic Relations

2.3.3.1 Conservation Equations

The conservation equations include mass, momentum, and energy conservation. According to the kinematic relations, the spatial description of the time derivative of any differentiable function (F) relative to the motion of a material point in phase (a) can be expressed as:

$$\frac{D^a F}{Dt} = \frac{\partial F}{\partial t} + v^a \nabla F \tag{2.19}$$

If phase (b) is taken as the reference phase, then:

$$\frac{D^b F}{Dt} = \frac{\partial F}{\partial t} + v^b \nabla F \tag{2.20}$$

By subtracting the two equations, we get:

$$\frac{D^b F}{Dt} = \frac{D^a F}{Dt} + v^{ba} \nabla F \tag{2.21}$$

where $v^{ba} = v^b - v^a$ is the velocity of phase (b) relative to phase (a)

1. Mass conservation equation

$$\frac{D^a \rho^a}{Dt} + \rho^a \nabla_i v_i^a = \bar{\rho}^a,\ a \in [S, L, G] \tag{2.22}$$

where, ρ^a is the apparent density of phase a; $\rho^a = n^a \rho_a$, where n^a is the volume-fraction of phase a, and $\sum n^a = 1$; ρ^a is the intrinsic density of phase a; $\bar{\rho}^a$ is the mass supply amount of other phases relative to phase a. For example, there is a one-way (solid →liquid) mass exchange term $\bar{\rho}_1$ between the solid and liquid phases due to temperature changes, and a mass exchange term $\bar{\rho}_2$ between the gas and liquid phases.

Define $\psi_{ij}^a = -\left(\nabla_i v_j^a + \nabla_j v_i^a\right)/2$ as the deformation rate of phase a, and stipulate that the material derivative with the solid phase as the reference phase can be expressed as $d_t(\cdot) = \partial_t(\cdot) + v_i^s \nabla_i(\cdot)$. Therefore, for the solid-phase soil skeleton,

$d_t\varepsilon_{ij} = \psi_{ij}^S - \varepsilon_{ik}\nabla_j v_k - \varepsilon_{jk}\nabla_i v_k$. Under the premise of the small-deformation assumption, the last two terms are very small in magnitude and can be ignored, then $d_t\varepsilon_{ij} \approx \psi_{ij}^S$.

$$\mathrm{d}_t\rho^S = \rho^S\psi_{kk}^S - \bar{\rho}_1 \tag{2.23a}$$

$$\mathrm{d}_t\rho^L = \rho^L\psi_{kk}^L - v_i^{LS}\nabla\rho^L + \bar{\rho}_1 - \bar{\rho}_2 \tag{2.23b}$$

$$\mathrm{d}_t\rho^G = \rho^G\psi_{kk}^G - v_i^{GS}\nabla\rho^G + \bar{\rho}_2 \tag{2.23c}$$

2. Momentum conservation equation

The momentum conservation equations for each phase of unsaturated soil can be expressed as:

$$\rho^a\frac{D^a v^a}{Dt} + \nabla \mathrm{n}^a\sigma^a - \rho^a b^a = \bar{\rho}^a v^a,\ a \in [S,L,G] \tag{2.24}$$

where, σ^a is the Cauchy stress tensor of phase a. According to the convention in soil mechanics, pressure is positive; b^a is the body-force vector of phase a. Here, $b^a = g$ where g is the acceleration due to gravity; $\bar{\rho}^a v^a$ is the contribution of mass exchange to the momentum of phase a.

According to de Boer [49], it is known that $\sum \bar{\rho}^a v_i^a \approx 0$, thus the overall momentum conservation equation for unsaturated soil can be expressed as:

$$\sum_{\alpha=S,L,G}\left(\rho_\alpha^d\frac{d\mathbf{v}_i^\alpha}{dt} + \nabla_j\sigma_{ij}^\alpha - \rho_\alpha^g\mathbf{g}\right) + \rho^L v_k^{LS}\nabla_k\mathbf{v}_l^L + \rho^G v_k^{GS}\nabla_k\mathbf{v}_l^G = 0 \tag{2.25}$$

3. Energy conservation equation

The energy conservation equation can generally be expressed as:

$$\frac{d}{dt}\iiint_\Omega w dV + \iiint_{\partial\Omega} Q_k da_k = g\iiint_\Omega \rho v_k dV \tag{2.26}$$

where, Ω is an arbitrary spatial region; da_k is the area-element vector defined in the outer-normal direction of $\partial\Omega$; w is the total energy density, $w = n^s w_s + n^L w_L + n^G w_G$; Q_k is the energy flux, simply referred to as energy flow.

For the energy per unit mass without kinetic energy $\rho v^2/2$, its specific energy form is $e = w/\rho - v^2/2$. According to Gauss's theorem, the energy conservation equation can be expressed in terms of specific energy e:

$$\rho\frac{De}{Dt} = \sigma_{kj}\psi_{jk} + \nabla_k\varpi_k \tag{2.27}$$

where $\varpi_k = w v_k + \sigma_{kj} v_j - Q_k$, representing the energy flow associated with dissipative processes such as heat conduction.

Therefore, the total energy conservation equation for the three phases of unsaturated soil can be expressed as:

$$\sum_{\alpha=S,L,G}\left(\rho_\alpha^d\frac{de_\alpha}{dt}\right)=\sum_{\alpha=S,L,G}\left(n^\alpha\sigma_{jk}^\alpha v_{jk}^\alpha\right)+\nabla_k\overline{q_k}-\rho^L v_k^{LS}\nabla_k e_L-\rho^G v_k^{GS}\nabla_k e_G \tag{2.28}$$

2.3.3.2 Entropy-Increase Equations

Entropy, like the aforementioned conserved quantities, possesses additivity and indestructibility. However, it differs in that it can be generated, that is, entropy production or heat generation. Entropy production is an irreversible phenomenon. Strictly speaking, a thermodynamic equilibrium is a prerequisite for it, but in theoretical analysis, it is found that the concept and behavior of entropy are usually valid even when deviating from thermal equilibrium. Similar to the form of the conservation equations, the entropy-increase equation can be expressed as:

$$\partial_t S+\nabla_k\left(Sv_k-f_k\right)=R/T \tag{2.29}$$

where, Sv_k is the entropy flow accompanying the mass flow; f_k represents the entropy transfer generated by heat-conduction. In thermodynamics, it is customary to express the entropy generation as the heat generation per unit volume per unit time, that is, the heat-generation rate R divided by the temperature T, $R \geq 0$, which satisfies the second law of thermodynamics.

The entropy capacity (entropy per unit mass) can be expressed as $\vartheta = S/\rho$. Similarly, according to Gauss's theorem, Eq. (2.29) can be expressed as:

$$\rho T\frac{D\vartheta}{Dt}=R+T\nabla_k f_k \tag{2.30}$$

Then, the total entropy-increase equation for unsaturated soil can be expressed as:

$$\sum_{\alpha=S,L,G}\left(\rho_\alpha^d\frac{d\vartheta_\alpha}{dt}\right)=R/T+\nabla_k f_k-\rho^L v_k^{LS}\nabla_k\vartheta_L-\rho^G v_k^{GS}\nabla_k\vartheta_G \tag{2.31}$$

Similar to Eq. (2.29), the kinematic equation of particle entropy can also be given.

$$\partial_t S_g+\nabla_k\left(S_g v_k-f_k^g\right)=R_g/T_g-I_g \tag{2.32}$$

where, $R_g \geq 0$, $I_g \geq 0$. R_g is the generation of particle entropy; f_k^g is the heat-conduction at the particle level; I_g is the decay rate of particle entropy, representing the decrease in particle entropy per unit volume per unit time.

Similarly, if the heat-conduction at the particle level is not considered, Eq. (2.32) can be expressed in the form of the specific entropy of particles $\vartheta_g = S_g/\rho^S$ as:

$$\rho^s d_t\vartheta_g=R_g/T_g-I_g \tag{2.33}$$

2.3.3.3 Thermodynamic Relations

From the independent thermodynamic state variables, the total differential form of thermodynamics for unsaturated soil can be expressed as:

$$dw = \pi_{ij} d\varepsilon_{ij}^{e} + T_g dS_g + TdS + \sum_{\alpha=S,L,G} \left(\mu_{c\alpha} d\rho^{\alpha} + v_i^{\alpha} dm_i^{\alpha} \right) \tag{2.34}$$

According to the above-mentioned equation, some relationships can be obtained, such as $\pi_{ij} = \partial w / \partial \varepsilon_{ij}^{e}$, $T_g = \partial w / \partial S_g$, etc.

(1) Determination of the dissipative system

Since mass, momentum, and energy are conserved quantities, for a certain unit-volume Ω of unsaturated soil, its total mass M, total momentum K, and total energy W will remain unchanged. However, due to inhomogeneity, ρ, m, and w at different positions can be different and change with time. When the system reaches a thermodynamic equilibrium state, according to the principles of thermodynamics, the total entropy S_t will be in a maximum state, that is, the overall equilibrium state with $R = 0$. When S_t is not in the maximum state, the system will allow dissipative processes that increase S_t. Therefore, by analyzing which spatial configurations of ρ, m, and w can lead to the maximum of the total entropy, the equilibrium state of the system and the dissipative situation can be determined.

$$M = \iiint_{\Omega} \rho dV, \ \rho = \rho^{s} + \rho^{L} + \rho^{G} \tag{2.35a-b}$$

$$K = \iiint_{\Omega} m dV, \ m = \rho^{s} v_i^{s} + \rho^{L} v_i^{L} + \rho^{G} v_i^{G} \tag{2.36a-b}$$

$$W = \iiint_{\Omega} w dV, \ w = n^{s} w_s + n^{L} w_L + n^{G} w_G \tag{2.37a-b}$$

$$S_t = \iiint_{\Omega} S dV, \ S = n^{S} S_S + n^{L} S_L + n^{G} S_G \tag{2.38a-b}$$

According to the principle of mathematical variation, analyze the constraint conditions where M, K, and W are constants, and find that the first-order variation of the total entropy S_t with respect to ρ, m and w is equal to zero, and the second-order variation is less than zero, then the dissipative composition of the system can be obtained:

$$\delta S_t = 0, \ \delta^2 S_t < 0 \tag{2.39a-b}$$

If the second-order variation is zero, higher-order variations need to be further analyzed. In general, this situation does not occur in actual geotechnical materials, so it can be left undiscussed. For the first-order variation δS_t, Lagrange multipliers are introduced to remove the constraints, and by using the total differential relation of thermodynamics (2.34), the following can be obtained:

$$\delta S_t = \iiint_{\Omega} \begin{bmatrix} \left(\ell_w + \frac{1}{T}\right)\delta \mathrm{w} + \left(\ell_m - \frac{\mu_{cS}}{T}\right)\delta \rho^S \\ + \left(\ell_m - \frac{\mu_{cL}}{T}\right)\delta \rho^L + \left(\ell_m - \frac{\mu_{cG}}{T}\right)\delta \rho^G \\ + \left(\ell_k - \frac{v_i^S}{T}\right)\delta m^S + \left(\ell_k - \frac{v_i^L}{T}\right)\delta m^L \\ + \left(\ell_k - \frac{v_i^G}{T}\right)\delta m^G - \frac{\pi_{ij}}{T}\delta \varepsilon_{ij}^e - \frac{T_g}{T}\delta S_g \end{bmatrix} dV = 0 \tag{2.40}$$

where, ℓ_v, ℓ_m, ℓ_k are Lagrange multipliers, which are constants independent of the spatial position. Equation (2.40) must hold for any variations δw, $\delta \rho^s$, $\delta \rho^L$, $\delta \rho^G$, δm^S, δm^L, δm^G, $\delta \varepsilon_{ij}^e$, δS_g. This requires satisfying the following relationships:

1. $T = -\ell_v^{-1}$, that is, $\nabla_i T = 0$. Select $\nabla_i T$ as the dissipative force, which corresponds to heat-conduction.
2. $\mu_{cS} = \mu_{cL} = \mu_{cG} = -T\ell_m$. Due to the limitations of causality and stability and other reasons, $\nabla_i \mu_{ca} = 0$ will not give additional dissipative processes and can be ignored [76].
3. $v_i^s = v_i^L = v_i^G = T\ell_k$. This indicates that the deformation rates of each phase and the relative-motion velocities between each phase are all the driving forces that cause the system to deviate from the equilibrium state, that is, $\psi_{ij}^s = 0$, $\psi_{ij}^L = 0$, $\psi_{ij}^G = 0$. Select the deformation rate ψ_{ij}^α of each phase as the dissipative force.
4. $\pi_{ij} = 0$, $T_g = 0$. Select π_{ij}, T_g as the dissipative forces.

Therefore, the set of dissipative forces for unsaturated soil can be obtained as $\{\pi_{ij}, T_g, \nabla_i T, d_t \varepsilon_{ij}, \psi_{ij}^L, \psi_{ij}^G\}$, and the corresponding dissipative fluxes are $\{Y_{ij}, I_g, q_k / T, \sigma_{ij}^{vS}, \sigma_{ij}^{vL}, \sigma_{ij}^{vG}\}$. Where, $\sigma_{ij}^{v\alpha}\sigma_{ij}^{v\alpha}$ is the viscous stress of phase a; Y_{ij} is the dissipative flux corresponding to the effective stress π_{ij}.

In the seepage problem of porous media, the study of phenomenological relations generally expresses the average velocity of the fluid as a function of the hydraulic gradient ∇h or the pore-water pressure μ_l, such as $\mathrm{v} = -K' \nabla h$, $\mathrm{v} = -K' \nabla u_l$, etc., where K' is the permeability coefficient. In this way, it is also easier to regard the average velocity v as the dissipative flux, and the viscous resistance or the hydraulic-head gradient ∇h as the dissipative force. This method of selecting dissipative force and flux is quite different from that in the problem of viscous flow. Because the former relates the dissipative force to the force in mechanics and the dissipative flux to motion and deformation; while the latter is the opposite. In fact, in the seepage problem of porous media, the selection of force and flux in the former is appropriate. In this way, it is convenient to select the hydraulic head of the seepage field as the basic unknown in the problem analysis, which can correspond to the traditional seepage-treatment methods. Therefore, for the seepage dissipation of the liquid and gas phases, selecting the relative velocities $v_i^{L,S}, v_i^{G,S}$ as the dissipative fluxes and the viscous stresses $\sigma_{ij}^{vL}, \sigma_{ij}^{vG}$ as the dissipative forces is equivalent to selecting $\sigma_{ij}^{vL}, \sigma_{ij}^{vG}$ as the dissipative fluxes and the deformation rates $\psi_{kl}^{L}, \psi_{kl}^{G}$ as the dissipative forces. In this paper, the viscous stress is selected as the dissipative flux and the deformation rate is selected as the dissipative force.

The Onsager [50] relation shows that the entropy-production rate R can be expressed as the sum of the products of dissipative forces and dissipative fluxes.

$$R = Y_{ij}\pi_{ij} + I_g T_g + \frac{q_k}{T}\nabla_i T + \sigma_{ij}^{vS} d_t \varepsilon_{ij} + \sigma_{ij}^{vL}\psi_{ij}^{L} + \sigma_{ij}^{vG}\psi_{ij}^{G} \tag{2.41}$$

For single-component particulate materials, the rise and fall motion of particles is generally only excited by mechanical motion (force-action), and cannot be excited by thermal motion (heating). For soil containing water and gas, heating can also cause the phenomenon of particle reorganization [57–63]. For example, heating can cause the consolidation effect of soil. From a microscopic perspective, heating will cause the conversion of bound water to free water, increasing the permeability coefficient, and making it easier to consolidate under external loads, that is, the reorganization motion occurs between particles. In other words, the phase change between bound water and free water will also cause energy dissipation, and the driving force for this phase change is the temperature change rate d_tT. That is, d_tT is a driving force for the rise and fall motion of particles, so it should be attributed to the dissipative mechanism at the particle level.

In addition, the change in saturation will also cause the reorganization motion between soil particles [21, 24]. Analyzing the influence of saturation on the mutual motion of soil particles from a microscopic perspective [65], when the saturation decreases, the matric suction increases, the attractive force between particles enhances, and the trend of particle motion decreases. In addition, due to the decrease in water content, the distance between molecules of the gel substances also becomes smaller, the molecular attractive force increases, and the cementation effect between particles increases. Therefore, the decrease in saturation will enhance the anti-disturbance ability of the soil. When the saturation increases, the surface tension at the shrinkage membrane and the matric suction decrease with the increase in water content, and the attractive force between particles decreases. Moreover, the increase in water content will increase the distance between particles, reduce the cementation force between particles, and lower the frictional resistance between particles, making it easier to cause the relative motion between particles. Therefore, the rate of change of saturation d_ts_r is also a driving force for the rise and fall motion of particles, and should also be attributed to the dissipative mechanism at the mesoscopic level.

Therefore, for unsaturated soil, heating can cause the soil to consolidate (thermal consolidation), that is, the change in temperature will cause the reorganization motion between soil particles, thereby triggering dissipation at the mesoscopic level. d_tT can be regarded as a dissipative force, and assume that R is the corresponding dissipative flux. Similarly, the change in saturation will also lead to the mutual-motion between particles, triggering dissipation at the mesoscopic level, such as the "collapse" phenomenon. $d_r s_r$ can be regarded as a dissipative force, and the suction s is the corresponding dissipative flux. Therefore, considering the viscous effect at the mesoscopic level and the dissipation at the mesoscopic level caused by temperature and saturation changes, the particle entropy-production rate R_g can be obtained as:

$$R_g = \sigma_{ij}^{vG} d_t \varepsilon_{ij} + \Re T_g d_t T + s T_g d_t s_r \tag{2.42}$$

where, $RT_g d_t T$ represents the dissipation at the mesoscopic level caused by the influence of temperature on the soil; $sT_g d_t s_r$ represents the dissipation at the mesoscopic level caused by the change in saturation. When there is no motion between particles, $T_g = 0$, that is, no dissipation occurs, a $RT_g d_t T$ and $sT_g d_t s_r$ also disappear accordingly (here, a similar treatment method in Ref. [47] is adopted, multiplying by T_g to reflect this phenomenon).

In the non-equilibrium region not far from the equilibrium state, the dissipative flux can be expressed as a linear function of the dissipative force [51], which is expressed as follows:

$$\begin{bmatrix} Y_{ij} \\ I_g \\ q_i \\ \sigma_{ij}^{vS} \\ \sigma_{ij}^{vL} \\ \sigma_{ij}^{vG} \end{bmatrix} = \begin{bmatrix} \eta_{ijkl} & 0 & 0 & 0 & 0 & 0 \\ 0 & \gamma & 0 & 0 & 0 & 0 \\ 0 & 0 & \kappa_{ik} & 0 & 0 & 0 \\ 0 & 0 & 0 & \lambda_{ijkl}^{S} & 0 & 0 \\ 0 & 0 & 0 & 0 & \lambda_{ijkl}^{L} & 0 \\ 0 & 0 & 0 & 0 & 0 & \lambda_{ijkl}^{G} \end{bmatrix} \begin{bmatrix} \pi_{kl} \\ T_g \\ \nabla_k T \\ d_t \varepsilon_{kl} \\ \psi_{kl}^{L} \\ \psi_{kl}^{G} \end{bmatrix} \tag{2.43}$$

Similarly, the matrix in the middle is referred to as the migration-coefficient matrix, and the coefficients within it are the correlation coefficients between the corresponding dissipative fluxes and dissipative forces. This matrix must be a positive-definite one and comply with the Onsager reciprocity and Curie principles. Since there are no significant mutual-influence relationships among the dissipative terms in Eq. (2.43), the influence of the correlation coefficients is not taken into account in this paper for the time being.

Likewise, the dissipation at the particle level can also be expressed as:

$$P_S = \left(\rho^S\right)^2 \mu_S = \rho^S \mu_{cS} + v_i^S m_i^S + Tn^S s_S + n^S T_g S_g - n^S w_S \tag{2.44}$$

where, λ_{ijkl}^{g} is the migration coefficient at the particle level, and σ_{ij}^{vg} is the particle viscous stress. Similarly, when $T_g = 0$, no dissipation occurs at the particle level, and the viscous dissipation also vanishes accordingly.

(2) Thermodynamic identities

The total differential equation of the thermodynamics of unsaturated soil can be expressed in the form of specific energy *e*.

$$\sum_{\alpha=S,L,G}\left(\rho^{\alpha} d_t e_{\alpha}\right)=\pi_{ij} d_t \varepsilon_{ij}^{e}+T_g \rho^{S} d_t \vartheta_g+\sum_{\alpha=S,L,G}(\rho^{\alpha}\mu_{\alpha} d_t \rho^{\alpha}+T\rho^{\alpha} d_t \vartheta_{\alpha} \quad (2.45)$$

where, $\mu_a=\left(\mu_{ca}+T\vartheta_a+T_g\vartheta_g+v_i^a v_i^a/2-e_a\right)/\rho^a$ is the chemical potential. When and only when $a = S$, it contains the term $T_g\vartheta_g$.

The hydrodynamic pressure P_T, also known as the hydrostatic pressure of the fluid, can be expressed as

$$P_T=\sum_{\alpha=S,L,G}\left[\left(\rho^{\alpha}\right)^2\mu_{\alpha}\right]=\sum_{\alpha=S,L,G}\left(\rho^{\alpha}\mu_{c\alpha}+v_i^{\alpha}m_i^{\alpha}\right)+TS+n^S T_g S_g-w \quad (2.46)$$

For the liquid phase, $P_L=\left(\rho^L\right)^2\mu_L=\rho^L\mu_{cL}+v_i^L m_i^L+Tn^L S_L-n^L w_L$, which is the pore-water pressure; for the gas phase, $P_G=\left(\rho^G\right)^2\mu_G=\rho^G\mu_{cG}+v_i^G m_i^G+Tn^G S_G-n^G w_G$, which is the pore-air pressure; for the soil skeleton, $P_S=\left(\rho^S\right)^2\mu_S=\rho^S\mu_{cS}+v_i^S m_i^S+Tn^S s_S+n^S T_g S_g-n^S w_S$, representing the response of soil particles compressed isotropically by the pore fluid. Even if there is no contact between particles, it is a non-zero phase. Therefore, P_s and the effective stress are two different concepts [55].

If the soil undergoes completely elastic deformation, then $d_t\varepsilon_{ij}^e=d_t\varepsilon_{ij}-d_t\varepsilon_{ij}^p$. When plastic deformation occurs, the elastic strain can be expressed as the total strain minus the plastic strain, that is, $d_t\varepsilon_{ij}^e=d_t\varepsilon_{ij}-d_t\varepsilon_{ij}^p$. Substituting the three conservation Eqs. (2.23), (2.25), (2.28) and the two entropy-increase eqs. (2.31), (2.33) into Eq. (2.45), the expression for the entropy-production rate can be obtained as:

$$\begin{aligned}R+R_g=&\left[n^G\sigma_{ij}^G-\mu_G\left(\rho^G\right)^2\delta_{ij}\right]\psi_{ij}^G+\left(\rho^L\mu_L\nabla_k\rho^L+T\rho^L\nabla_k\vartheta_L-\rho^L\nabla_k e_L\right)v_k^{LS}\\&+\left(\rho^G\mu_G\nabla_k\rho^G+T\rho^G\nabla_k\vartheta_G-\rho^G\nabla_k e_G\right)v_k^{GS}+\pi_{ij}d_t\varepsilon_{ij}^p+T_g I_g+f_k\nabla_k T\\&+\left[n^S\sigma_{ij}^S-\pi_{ij}-\mu_S\left(\rho^S\right)^2\delta_{ij}\right]\psi_{ij}^S+\left[n^L\sigma_{ij}^L-\mu_L\left(\rho^L\right)^2\delta_{ij}\right]\psi_{ij}^L\\&+\nabla_k\left(\varpi_k-f_k T\right)+\left(\rho^S\mu_S-\rho^L\mu_L\right)\overline{p}_1+\left(\rho^L\mu_L-\rho^G\mu_G\right)\overline{p}_2\end{aligned} \quad (2.47)$$

Combining Eq. (2.41) and Eq. (2.42), the entropy-production rate can also be expressed as:

$$\begin{aligned}R+R_g=&Y_{ij}\pi_{ij}+I_g T_g+\frac{q_k}{T}\nabla_i T+\left(\sigma_{ij}^{vS}+\sigma_{ij}^{vg}\right)\psi_{ij}^{s}\\&+\sigma_{ij}^{vL}\psi_{ij}^L+\sigma_{ij}^{vG}\psi_{ij}^G+\Re T_g d_t T+sT_g d_t s_r\end{aligned} \quad (2.48)$$

Through the one-to-one correspondence between Eq. (2.47) and Eq. (2.48), the following relationships can be obtained:

$$\varpi_k - f_k T = 0,\ q_k = f_k T \tag{2.49a-b}$$

Therefore, $\varpi_k = q_k$ can be obtained, which verifies that ϖ_k in the energy-conservation Eq. (2.28) is the energy-flow contribution accompanied by dissipations such as heat conduction.

Similarly, it can also be concluded that:

$$Y_{ij} = d_t \varepsilon_{ij}^p \tag{2.50}$$

That is, the rate of change of plastic strain is the dissipative flux corresponding to the effective stress π_{ij}.

The expressions related to the stresses of the solid, liquid, and gas phases are:

$$n^S \sigma_{ij}^S = \pi_{ij} + \mu_S \left(\rho^S\right)^2 \delta_{ij} + \sigma_{ij}^{vS} + \sigma_{ij}^{vg} = \pi_{ij} + p_S \delta_{ij} + \sigma_{ij}^{vS} + \sigma_{ij}^{vg} \tag{2.51}$$

$$n^L \sigma_{ij}^L = \mu_L \left(\rho^L\right)^2 \delta_{ij} + \sigma_{ij}^{vL} = p_L \delta_{ij} + \sigma_{ij}^{vL} \tag{2.52}$$

$$n^G \sigma_{ij}^G = \mu_G \left(\rho^G\right)^2 \delta_{ij} + \sigma_{ij}^{vG} = p_G \delta_{ij} + \sigma_{ij}^{vG} \tag{2.53}$$

For the liquid and gas phases, since $de = Td\vartheta + \mu d\rho$, dividing it by the spatial interval ∂x_i gives:

$$\nabla_k e = T \nabla_k \vartheta + \mu \nabla_k \rho \tag{2.54}$$

Therefore, it can be concluded that:

$$\rho^L \nabla_k e_L = \rho^L T \nabla_k \vartheta_L + \rho^L \mu_L \nabla_k \rho^L \tag{2.55}$$

$$\rho^\alpha \nabla_k e_\alpha = \rho^G T \nabla_k \vartheta_G + \rho^G \mu_G \nabla_k \rho^G \tag{2.56}$$

Therefore, in Eq. (2.47), the values inside the parentheses in front of v_i^{LS}, v_i^{GS} are zero.

In addition, based on the identity relationships, it can also be concluded that:

$$\left(\rho^S \mu_S - \rho^L \mu_L\right) \overline{\rho}_1 = \Re T_g d_t T \tag{2.57}$$

$$\left(\rho^L \mu_L - \rho^G \mu_G\right) \overline{\rho}_2 = \mathrm{s} T_g d_t s_r \tag{2.58}$$

As known from the previous text, $\overline{\rho}_1$ is the conversion amount between the solid (bound water) and the liquid phases caused by the temperature change, so there is:

$$\overline{\rho}_1 = \rho^L \alpha_{LT} \mathrm{d}_t T \tag{2.59}$$

where, α_{LT} is the rate of conversion of bound water into free pore-water during heating, and it can be taken as 0.002 °C^{-1}.

According to Eq. (2.46), it is known that:

$$\rho^S \mu_S = P_S / \rho^S \tag{2.60a}$$

$$\rho^L \mu_L = P_L / \rho^L \tag{2.60b}$$

$$\rho^G \mu_G = P_G / \rho^G \tag{2.60c}$$

Therefore, substituting Eq. (2.59) and Eq. (2.60) into Eq. (2.57), it can be concluded that:

$$\Re T_g = \left(\rho^S \mu_S - \rho^L \mu_L\right)\bar{\rho}_1 = \left(P_S / \rho^S - u_l / \rho^L\right)\rho^L \alpha_{LT} \approx \alpha_{LT}\alpha_v p' \tag{2.61}$$

where, p' represents the average effective stress; α_v is the proportionality coefficient.

Since heating causes the conversion between bound water and free water, the amount of bound water decreases, and the bound-water film layer becomes thinner. As a result, the stress state between particles changes, the phenomenon of particle reorganization occurs, and dissipation at the particle level is generated. This also shows that RT_g is related to the dissipation at the particle level caused by temperature. Therefore, P_s/ρ^s in Eq. (2.61) is strictly for the bound-water phase. And because the model assumes that the bound-water phase is classified as a solid-phase component, α_v can be regarded as the proportion of the effective stress borne by the bound water.

In addition, $\bar{\rho}_2$ is the mass-conversion amount between the liquid and gas phases, which is mainly caused by temperature changes. The changes between the liquid and gas phases will in turn cause changes in the degree of saturation, and the changes in the degree of saturation are also related to temperature. Therefore, Eq. (2.58) can be regarded as reflecting the conversion relationship between the liquid and gas phases through the change in the degree of saturation. It can also be seen from Eq. (2.60) that the term $(\rho^L \mu_L - \rho^G \mu_G)$ is related to the suction s.

2.3.4 *Potential-Energy Density Function*

2.3.4.1 Elastic-Potential-Energy Density Function

The elastic-potential-energy density function is generally expressed as a function of elastic strain. For example, in the well-known linear elastic model (generalized Hooke's law), some scholars [41, 66, 67] determine the elastic-potential-energy density function according to the influence of the average effective stress on the elastic bulk modulus and the elastic shear modulus, and, similar to the form of the linear elastic model, express it as a function of elastic strain. Jiang and Liu [41] gave the elastic-potential-energy density function of particle-like materials as follows by making an analogy with Hertzian contact, regarding the contacting spheres as particle solids and the springs as Hookean solids.

$$w_e = B\left(\varepsilon_v^e\right)^{0.5}\left[\frac{2}{5}\left(\varepsilon_v^e\right)^2 + \frac{1}{\xi}\left(\varepsilon_z^e\right)^2\right] \tag{2.62}$$

The initial form of Eq. (2.62) is $w_e = \frac{1}{2}\left(\varepsilon_v^e\right)^{0.5}\left[B\left(\varepsilon_v^e\right)^2 + A\left(\varepsilon_z^e\right)^2\right]$. In order to ensure energy stability, the cross-convexity condition needs to be satisfied, that is, $\left(\varepsilon_s^e\right)^2/\left(\varepsilon_v^e\right)^2 \le 5B/2A$. For the convenience of representation, multiplying the term $\left(\varepsilon_s^e\right)^2/\left(\varepsilon_v^e\right)^2 \le 5B/2A$ by 2/5 can make it expressed as $\left(\varepsilon_s^e\right)^2/\left(\varepsilon_v^e\right)^2 \le B/A$. Therefore, 2/5 is a factor introduced for the convenience of representation [68]. It should be noted that Eq. (2.62) is a conjecture proposed by analogy with Hertzian contact and is not obtained through strict mathematical derivation. Hertz first analyzed the difference in elastic properties between a common spring and a two-solid-sphere contact system. The former satisfies Hooke's law, and the elastic potential energy is $w_e = k\left(\varepsilon_v^e\right)^2/2$, where k is the elastic coefficient. He found that the contact of solid spheres is equivalent to a non-linear spring with an elastic coefficient $k \sim \sqrt{\varepsilon_v^e w_e}$, that is, the elastic potential energy of the sphere contact system is $w_e \sim \sqrt{\varepsilon_v^e}\left(\varepsilon_v^e\right)^2$. If the spring is regarded as a Hookean solid with $w_e \sim \left(\varepsilon_v^e\right)^2 + \frac{1}{\xi}\left(\varepsilon_s^e\right)^2$ and the Hertzian contact spheres are regarded as particle solids, it can be inferred that the elastic potential energy of the particle solid should be the product of $\left[\left(\varepsilon_v^e\right)^2 + \frac{1}{\xi}\left(\varepsilon_z^e\right)^2\right]$ and $\sqrt{\varepsilon_v^e}$, thus obtaining Eq. (2.62). Although this equation is an assumption, it can qualitatively reflect some mechanical properties of particulate materials [68, 69].

Zhang [48] considered the case of saturated clay and believed that both the elastic bulk modulus and the elastic shear modulus are affected by temperature, and the elastic shear modulus is also affected by cohesion. A material parameter c was introduced to reflect the influence of viscosity on the elastic shear modulus, $G_e \propto \left(\varepsilon_v^e + \beta_T \Delta T + c\right)^{0.5}$. Based on Eq. (2.62), the influence of cohesion and temperature was taken into account, and the following modification was made:

$$w_e = B\left[\left(\varepsilon_v^e + \beta_T \Delta T\right)^{2.5} + \zeta\left(\varepsilon_v^e + \beta_T \Delta T + c\right)^{0.5}\left(\varepsilon_z^e\right)^2\right] \tag{2.63}$$

where, c is a material parameter reflecting the strength of viscosity; ζ is a material parameter similar to $1/\xi$ in Eq. (2.62), so it can be approximately taken as 0.6; β_T is the linear expansion coefficient of soil, generally taken as 1×10^{-5} °C^{-1}.

For unsaturated soils, in addition to being affected by temperature and viscosity, there is also the influence of matric suction. By following the method of describing the thermal expansion and contraction effect of solids in Ref. [70], a term $\chi s \varepsilon_v^e$ describing the contribution of matric suction to elastic potential energy is added to Eq. (2.63). Then, the elastic-potential-energy density function for unsaturated soils can be modified as:

$$w_e = B\left[\frac{2}{5}\left(\varepsilon_v^e + \beta_T \Delta T\right)^{2.5} + \zeta\left(\varepsilon_v^e + \beta_T \Delta T + c\right)^{0.5}\left(\varepsilon_z^e\right)^2\right] + \chi s \varepsilon_v^e \quad (2.64)$$

The effective stress of unsaturated soil in the form of Bishop [71] can be expressed as $\sigma_{ij'} = \sigma_{ij} - P_T \delta_{ij}$, where $P_T = P_a - \chi(P_a - P_L)$. Many scholars [72–74] replaced s_r with the degree of saturation χ, which is also called the simplified Bishop formula. Here, in a similar way, let χ in Eq. (2.64) be replaced by the degree of saturation s_r.

2.3.4.2 The Expression of Stress

The energy density w can be written as the sum of the contributions of several energies as:

$$w = w_0 + w_e + w_g + w_k \quad (2.65)$$

where, w_0 is the thermal energy of the system when it is at rest and not disturbed by the outside world; w_e is the elastic potential energy stored when the system deforms under force; w_g is the energy related to the random fluctuations and movements of particles; w_k is the kinetic energy.

As known from Sect. 2.3.3.3, $\pi_{ij} = \partial W / \partial \varepsilon_{ij}^e$, and the energy density related to elastic strain is mainly provided by the elastic-potential-energy density. Therefore, $\pi_{ij} = \partial w_e / \partial \varepsilon_{ij}^e$. Substituting Eq. (2.64) into it, the expression of π_{ij} can be obtained:

$$\pi_{ij} = \frac{\partial w_e}{\partial \varepsilon_{ij}^e} = B\begin{bmatrix}\left(\varepsilon_v^e + \beta_T \Delta T\right)^{1.5}\delta_{ij} + 0.5\zeta\left(\varepsilon_v^e + \beta_T \Delta T + c\right)^{-0.5} \\ \left(\varepsilon_z^e\right)^2\delta_{ij} + 2\zeta\left(\varepsilon_v^e + \beta_T \Delta T + c\right)^{0.5} e_{ij}^e\end{bmatrix} + s s_r \delta_{ij} \quad (2.66)$$

Since under the constraint of external forces, thermal expansion will cause interactions between soil particles, temperature will have an impact on both the elastic bulk modulus K_e and the elastic shear modulus G_e. Referring to the forms in Zhang [48], we have $K_e \propto \left(\varepsilon_v^e + \beta_T \Delta T\right)^{0.5}$ and $G_e \propto \left(\varepsilon_v^e + \beta_T \Delta T + c\right)^{0.5}$. Therefore, the elastic bulk modulus K_e and the elastic shear modulus G_e considering the influence of temperature can be defined respectively as:

$$K_e = B\left[\left(\varepsilon_v^e + \beta_T \Delta T\right)^{0.5} + 0.5\zeta\left(\varepsilon_v^e + \beta_T \Delta T + c\right)^{-0.5}\frac{\left(\varepsilon_z^e\right)^2}{\varepsilon_v^e + \beta_T \Delta T}\right] \quad (2.67)$$

$$G_e = B\zeta\left(\varepsilon_v^e + \beta_T \Delta T + c\right)^{0.5} \quad (2.68)$$

Similar to the form of the Cauchy elastic model, π_{ij} can also be expressed as:

$$\pi_{ij} = \left[K_e\left(\varepsilon_v^e + \beta_T \Delta T\right) + s s_r\right]\delta_{ij} + 2G_e e_{ij}^e \quad (2.69)$$

In addition, the average effective stress of unsaturated soil can be expressed as:

$$p' = \frac{\pi_{kk}}{3} = B\left[\left(\varepsilon_v^e + \beta_T \Delta T\right)^{1.5} + 0.5\zeta\left(\varepsilon_v^e + \beta_T \Delta T + c\right)^{-0.5}\left(\varepsilon_s^e\right)^2\right] + ss_r \quad (2.70)$$

Under isothermal conditions ($\Delta T = 0$), when the critical state is reached, ε_v^e, ε_s^e remain unchanged, and the average net stress $\overline{p}$ can be expressed as:

$$\overline{p} = p' - ss_r = B\left[\left(\varepsilon_v^e\right)^{1.5} + 0.5\zeta\left(\varepsilon_v^e + c\right)^{-0.5}\left(\varepsilon_s^e\right)^2\right] \quad (2.71)$$

The test results show that, similar to saturated soil, when the critical state is reached, the void ratio and the average net stress of unsaturated soil under different suctions also approximately exhibit a linear relationship in a semi-logarithmic coordinate system [75].

$$\Gamma = \mathrm{e} + \lambda \ln \overline{p} \quad (2.72)$$

where, λ is the slope of the compression curve, which is related to the suction s and can be expressed as a function of the suction [29] $\lambda(\mathrm{s}) = \lambda(0)[(1 - r)\exp(-\beta s) + r]$, where $\lambda(0)$ is the slope of the saturated-soil compression curve, and$r$, β are fitting parameters.

Taking the logarithm of Eq. (2.71), we can get

$$\ln \overline{p} = \ln B + \ln \mathrm{A} \quad (2.73)$$

where, $A = \left(\varepsilon_v^e\right)^{1.5} + 0.5\zeta\left(\varepsilon_v^e + c\right)^{-0.5}\left(\varepsilon_s^e\right)^2$, which can be regarded as a constant under the critical state.

It is assumed that B can be expressed in the form of $B_0 \exp(\Xi)$. B_0 can be regarded as the bulk elastic modulus of soil particles. Substituting it into Eq. (2.73), we can get:

$$\ln p' = \Xi + \ln B_0 + \ln A \quad (2.74)$$

Corresponding to Eq. (2.72), it can be approximately deduced that $\Xi = -e/\lambda$. The specific volume v can also be used to replace the void ratio e, where $v = e + 1$. Therefore, B can be approximately expressed as:

$$B = B_0 \exp\left(-\frac{v-1}{\lambda(s)}\right) \quad (2.75)$$

It is known that B is positively correlated with the dry density ρ_d, and the dry density ρ_d is related to the void ratio e [41].

$$e = \frac{\rho_L G_s}{\rho_d} - 1 \quad (2.76)$$

Therefore, the obtained form of B (Eq. (2.75)) can establish a relationship with the compactness of the soil. Its value is jointly determined by B_0, e and $\lambda(s)$. When the suction increases, the compactness of the soil usually increases, and thus the value of B will also increase.

2.3.5 Migration-Coefficient Model

1. Soil-skeleton strain

According to the relationship between the dissipative flux and the dissipative force (Eq. (2.43)) and the thermodynamic identity (Eq. (2.50)), we can obtain:

$$d_t\varepsilon_{ij}^{p} = \eta_{ijkl}\pi_{kl} \tag{2.77}$$

When the particle temperature T_g is zero, the plastic-strain rate no longer develops. Therefore, the plastic-strain rate should be a function of T_g. In the case of isotropy, the migration coefficient η_{ijkl} can be expressed as:

$$\eta_{ijkl} = \left(T_g\right)^{\alpha}\frac{\eta_s}{2G_e}\delta_{il}\delta_{jk} + \left(T_g\right)^{\alpha}\left(\frac{\eta_v}{3K_e} - \frac{\eta_s}{6G_e}\right)\delta_{ij}\delta_{kl} \tag{2.78}$$

Among them, η_v and η_s are migration coefficients, which are a representation of the migration-coefficient tensor η_{ijkl}. Assuming that the elastic shear strain in the critical state is independent of the shear-rate, referring to Ref. [47], a can be taken as 0.5.

Substitute Eq. (2.78) into Eq. (2.77), and define $c_1 = \eta_v/\eta_s$. Then the plastic-strain rate can be expressed as:

$$d_t\varepsilon_{ij}^{p} = \eta_{ijkl}\pi_{kl} = c_1\left(\eta_s^{1/\alpha}T_g\right)^{0.5}\left(\varepsilon_v^e + \beta_T\Delta T + \frac{ss_r}{K_e}\right)\delta_{ij} + \left(\eta_s^{1/\alpha}T_g\right)^{0.5}e_{ij}^e \tag{2.79}$$

Therefore, the total strain-rate can be expressed as:

$$d_t\varepsilon_{ij} = d_t\varepsilon_{ij}^{e} + c_1\left(\eta_s^{1/\alpha}T_g\right)^{\alpha}\left(\varepsilon_v^e + \beta_T\Delta T + \frac{ss_r}{K_e}\right)\delta_{ij} + \left(\eta_s^{1/\alpha}T_g\right)^{\alpha}e_{ij}^e \tag{2.80}$$

2. Granular temperature motion equation.

Jiang and Liu [41] pointed out that the energy w_g related to the fluctuating motion of particles can be expressed as:

$$\mathrm{w}_g = b\rho^S\left(T_g\right)^2/2 \tag{2.81}$$

where, b is a material parameter.

According to the non-saturated soil thermodynamics differential eq. (2.45), it is known that:

$$\rho^s \vartheta_s = \frac{\partial w_g}{\partial T_g} = \rho^s bT_g \tag{2.82}$$

Thus, $\vartheta_g = bT_g$. Similar to Eq. (2.78), under isotropic conditions, the particle-level migration coefficient $\lambda_{ijkl}{}^g$ can be expressed as:

$$\lambda_{ijkl}^g = \lambda_{sg} T_g \delta_{il} \delta_{jk} + \left(\lambda_{vg} - \frac{\lambda_{sg}}{3} \right) T_g \delta_{ij} \delta_{lk} \tag{2.83}$$

where, λ_{vg} and λ_{sg} are the migration coefficients at the particle level, and this formula is a representation of the tensor-form migration coefficientλ_{ijkl}^g.

Combining Eqs. (2.33), (2.42), (2.44), (2.82), and (2.83), the motion equation of the particle temperature can be obtained:

$$d_t\left(T_{gg}\right) = c_2 c_5 \frac{d_t e_{ij} d_t e_{ij}}{\rho^s} + c_3 c_5 \frac{d_t \varepsilon_v d_t \varepsilon_v}{\rho^s} + c_4 \frac{\Re d_t T + s d_t s_r}{\rho^s} - c_5 \frac{T_{gg}}{\rho^s} \tag{2.84}$$

where, $c_2 = \eta_s^{1/\alpha} \lambda_{sg} / \gamma$, $c_3 = \eta_s^{1/\alpha} \lambda_{vg} / \gamma$, $c_4 = \eta_s^{1/\alpha} / b$, $c_5 = \gamma / b$, $T_{gg} = \eta_s^{1/\alpha} T$.

3. Some other relationships

$$q_i / T = \kappa_{ij} \nabla_j T \tag{2.85}$$

This is the heat-conduction equation. In the case of isotropy, $\kappa_{ij} = \kappa\delta_{ij}$.
In addition, the following relationships also exist:

$$\sigma_{ij}^{vS} = \lambda_{ijkl}^S d_t \varepsilon_{kl}, \sigma_{ij}^{vL} = \lambda_{ijkl}^L \psi_{kl}^L, \sigma_{ij}^{vG} = \lambda_{ijkl}^G \psi_{kl}^G, I_g = \gamma T_g \tag{2.86a–d}$$

2.3.6 *Soil-Water Characteristic*

The soil-water characteristic curve generally adopts the van Genuchten [31] model (vG model), that is,

$$s_r = \left[1 + \left(\frac{s}{a} \right)^n \right]^{-m} \tag{2.87}$$

where, a is generally considered to be related to the air-entry value, n is related to the slope of the soil-water characteristic curve ($s_r \sim$ ln s), and m is related to the asymmetry of the curve. All three are fitting parameters. For the drying and wetting curves, the values of a, m, and n are also different.

Experimental studies have found that both deformation and temperature can affect the soil-water characteristic curve (SWCC). Some scholars [64, 77, 78] proposed SWCC models considering the influence of deformation or temperature on the basis of the vG model. Gallipoli [78] obtained the SWCC considering the influence of deformation through experimental fitting on the basis of the vG model, and its form is as follows:

$$s_r = \left[1 + (as)^n\right]^{-m} \tag{2.88}$$

where, $a = \phi(v - 1)^{\varphi}$, n, m, ϕ, φ can be obtained through three-dimensional fitting.

Tarantion [64] found in experimental studies that the pore water content e_w and the suction s exhibit a power-function relationship, $e_w = a_1 s^{b1}$, especially in the high-suction range. Among them, a_1 and b_1 are related to the intercept and slope of the ln$s$~ ln e_w relationship curve, respectively. Similarly, on the basis of the vG model, the expression of the SWCC considering the influence of deformation is given as:

$$s_r = \left\{1 + \left[\left(\frac{e}{a_1}\right)^{1/\beta_1} s\right]^n\right\}^{-b_1/n} \tag{2.89}$$

where, a_1, b_1, and n are all fitting parameters; $b_1/n = m$, which is the same as m in the vG model.

Since the states below the main wetting surface or above the main drying surface are states that the soil cannot reach, if the soil state moves along the main wetting/drying surface, irreversible changes in saturation will occur. When the change in the void ratio is not considered, Eq. (2.89) is the same as the vG model. The main drying and wetting curves are shown in Fig. 2.2. When the change in the void ratio is considered, the spatial relationship among e ~ ln s ~ s_r is shown in Fig. 2.3. When

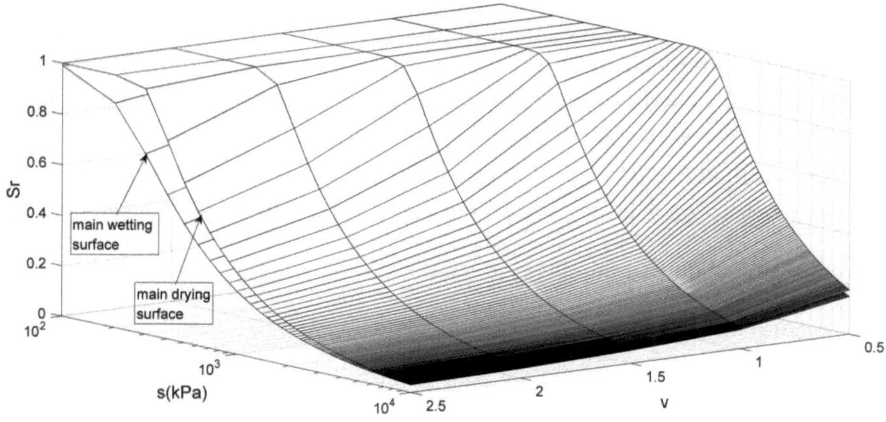

Fig. 2.2 Main drying and wetting curves without considering the change of void ratio

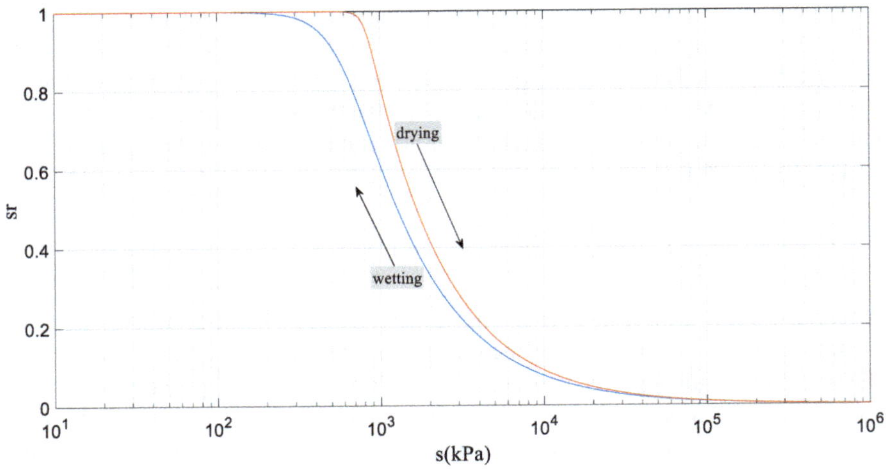

Fig. 2.3 Typical main drying and main wetting surfaces of Speswhite kaolin in $e \sim lns \sim s_r$ space

the suction is constant, as the volume v increases, the saturation will decrease; when the saturation is constant, as the volume v increases, the suction will also increase. It can be seen that the change in the void ratio will have a certain impact on the SWCC.

Grant and Salehzadeh [77] considered the influence of temperature on the SWCC and modified the vG model on the basis of the vG model as follows:

$$s_r = \left[1 + \left(\frac{b_2 + T_0}{b_2 + T}\frac{s}{a}\right)^n\right]^{-m} \tag{2.90}$$

where, $b_2 = a'/b' = -493.5^\circ\mathrm{C}$ is a constant, and a' and b' are fitting parameters. The relationship between the surface tension of the water-air interface and temperature can be approximately expressed as $\sigma_T = a' + b'T$, where σ_T is the surface tension. Grant and Salehzadeh obtained through parameter calibration that for the interface between pure water and air, $a' = 0.1177\mathrm{N \cdot m^{-1}}$. Since the value of b_2 is calibrated according to the interface between pure water and air, the b_2 for different types of soil can also be approximately regarded as the same.

It is not difficult to find that both Eq. (2.89) and Eq. (2.90) are obtained by introducing relevant influence factors on the basis of the vG model. Equation (2.89) introduces the deformation influence factor $\left(\frac{e}{a_1}\right)^{1/b_1}$, and Eq. (2.90) introduces the temperature influence factor $\frac{b_2 + T_0}{b_2 + T}$. By combining Eq. (2.89) and Eq. (2.90) and introducing the influence factors of deformation and temperature at the same time, the soil-water characteristic curve can be expressed as:

$$s_r = \left\{ 1 + \left[\frac{b_2 + T_0}{b_2 + T} \left(\frac{e}{a_1} \right)^{1/b_1} s \right]^n \right\}^{-b_1/n} \tag{2.91}$$

Equation (2.91) is the SWCC model that simultaneously considers the influences of deformation and temperature. The model is used to fit the experimental data of Uchaipichat and Khalili [79], and the results are shown in Fig. 2.4. It can be seen that the improved SWCC model can well simulate the changes of the soil-water characteristic curve affected by temperature and deformation.

2.3.7 Shear Characteristics

Different from the consolidation characteristics, the shear-strength of unsaturated soil can be considered to be related only to a single stress-state variable [71]. Some scholars [80, 81] believe that the effective stress in the form of Bishop is sufficient to describe the changes in the shear-strength of some soils, as shown in Fig. 2.5. It can be seen that for soils in different suction states, their critical-state lines are basically the same [30].

The shear stress of the model proposed in this chapter can be expressed as:

$$q = \sqrt{\frac{3}{2} s_{ij} s_{ji}} = \sqrt{6} B\zeta \left(\varepsilon_v^e + \beta_T \Delta T + c \right)^{0.5} \varepsilon_s^e \tag{2.92}$$

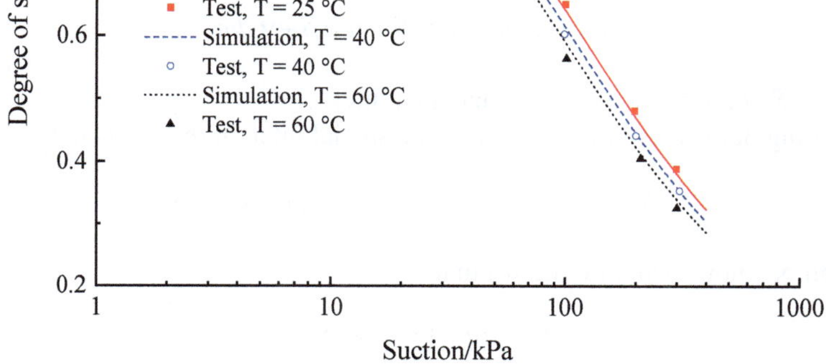

Fig. 2.4 Simulation results of SWCC at different temperatures (p0 = 150 kPa)

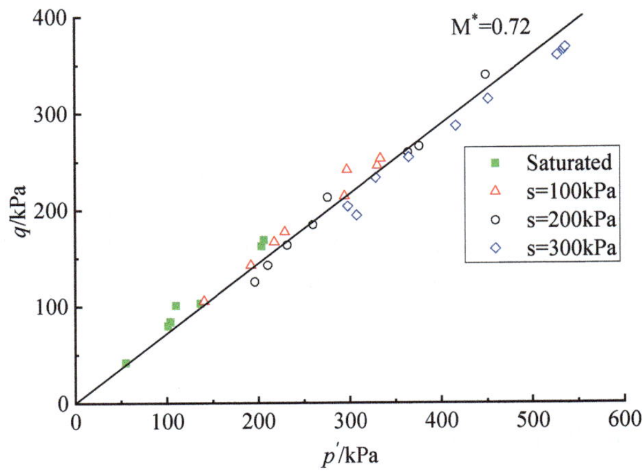

Fig. 2.5 Critical state line of Speswhite kaolin [228]

where, $s_{ij} = \pi_{ij} - \frac{\pi_{kk}}{3}\delta_{ij}$ is the deviatoric stress.

2.3.8 The Effective Stress of Unsaturated Soil

Bishop [71] expressed the form of the effective stress of unsaturated soil as:

$$\sigma_{ij'} = \sigma_{ij} - P_t\delta_{ij} \tag{2.93}$$

where, $P_T = u_G - \chi(u_G - u_L)$, and u_G are the pore-air pressure and pore-water pressure respectively.

From the above content, it can be known that when in an equilibrium state, $T_g \equiv 0$, and the average total differential equation of the three-phase mixture of unsaturated soil can be expressed as:

$$dw = \pi_{ij} d\varepsilon_{ij}^{e} + TdS + \mu_c d\rho + v_i dm \tag{2.94}$$

where, T, S, μ_c, ρ, v_i, m are all average quantities.

Dividing both sides of Eq. (2.94) by the spatial interval ∂x_k, we can get:

$$\nabla_k w = \pi_{ij} \nabla_k \varepsilon_{ij}^{e} + T\nabla_k S + \mu_c \nabla_k \rho + v_i \nabla_k m \tag{2.95}$$

It can be known from Eq. (2.46) that:

$$P_T = \rho\mu_c + v_i m + TS - w \tag{2.96}$$

Similarly, divide both sides of Eq. (2.96) by the spatial interval ∂x_k and substitute it into Eq. (2.95) to obtain:

$$\nabla_k P_T = \rho \nabla_k \mu_c + m \nabla_k v_k + S \nabla_k T - \pi_{ij} \nabla_k \varepsilon_{ij}^e \tag{2.97}$$

Dividing both sides of Eq. (2.94) by the time-interval ∂_t, we can obtain

$$\partial_t w = \pi_{ij} \partial_t \varepsilon_{ij}^e + T \partial_t S + \mu_c \partial_t \rho + v_i \partial_t m \tag{2.98}$$

In the equilibrium state, ignoring the influence of body forces, the entropy-increase equation and the three conservation equations of mass, momentum, and energy can be expressed respectively as:

$$\partial_t S + \nabla_k (S v_k) = 0 \tag{2.99}$$

$$\partial_t \rho + \nabla_k (\rho v_k) = 0 \tag{2.100}$$

$$\partial_t m + \nabla_k (\sigma_{ik} + \rho v_k v_i) = 0 \tag{2.101}$$

$$\partial_t w + \nabla_k Q_k = 0 \tag{2.102}$$

According to Eqs. (2.99), (2.100), and (2.101), we can obtain respectively:

$$-T \partial_t S = T \nabla_k (S v_k) = \nabla_k (T S v_k) - S v_k \nabla_k T \tag{2.103}$$

$$-\mu_c \partial_t \rho = \mu_c \nabla_k (\rho v_k) = \nabla_k (\mu_c \rho v_k) - \rho v_k \nabla_k \mu_c \tag{2.104}$$

$$\begin{aligned} -v_i \partial_t m &= v_i \nabla_k (\sigma_{ik} + \rho v_k v_i) = v_i \nabla_k \left(\sigma_{ik'} + P_T \delta_{ij} + \rho v_k v_i\right) \\ &= \nabla_k \left(\sigma_{ik'} v_i\right) - \sigma_{ik'} \nabla_k v_i + \nabla_k (\rho v_k v_i v_i) - \rho v_k v_i \nabla_k v_i + v_i \nabla_k P_T \end{aligned} \tag{2.105}$$

Substitute Eqs. (2.102), (2.103), (2.104), and (2.105) into Eq. (2.98) to obtain:

$$\begin{aligned} \nabla_k Q_k &= -\partial_t w = \nabla_k (T s v_k) - S v_k \nabla_k T + \nabla_k (\mu_c \rho v_k) - \rho v_k \nabla_k \mu_c \\ &+ \nabla_k \left(\sigma'_{ik} v_i\right) - \sigma'_{ik} \nabla_k v_i + \nabla_k (\rho v_k v_i v_i) - \rho v_k v_i \nabla_k v_i + v_i \nabla_k P_T \end{aligned} \tag{2.106}$$

Substitute Eq. (2.97) into Eq. (2.106) to obtain:

$$\nabla_k Q_k = \nabla_k \left[(TS + \mu_c \rho + \rho v_i v_i) v_k + \sigma'_{ik} v_i \right] - \pi_{ij} \partial_t \varepsilon_{ij}^e - v_i \pi_{ij} \nabla_k \varepsilon_{ij}^e - \sigma'_{ik} \nabla_k v_i \tag{2.107}$$

Also, since $T_g \equiv 0$ and $d_t \varepsilon_{ik}^e = \psi_{ik}^S$, it can be known that:

$$-\pi_{ik} \partial_t \varepsilon_{ik}^e - v_i \pi_{ik} \nabla_k \varepsilon_{ik}^e = -\pi_{ik} d_t \varepsilon_{ik}^e = -\pi_{ik} \psi_{ik}^S \tag{2.108}$$

Since $\sigma'_{ik} = \sigma'_{ki}$, we also have:

$$-\sigma'_{ik} \nabla_k v_i = -\frac{1}{2} \left(\sigma'_{ik} \nabla_k v_i + \sigma'_{ki} \nabla_i v_k \right) = \sigma'_{ik} \psi_{ik}^S \tag{2.109}$$

So

$$\nabla_k Q_k = \nabla_k \left[\left(TS + \mu_c \rho + \rho v_i v_i \right) v_k + \sigma'_{ik} v_i \right] + \left(\sigma'_{ik} - \pi_{ik} \right) \psi^S_{ik} \tag{2.110}$$

Since Eq. (2.110) needs to hold for any ψ^s_{ij}, we have

$$\sigma'_{ik} = \pi_{ik} \tag{2.111}$$

$$Q_k = \left(TS + \mu_c \rho + \rho v_i v_i \right) v_k + \sigma'_{ik} v_i \tag{2.112}$$

The above-mentioned proof shows that π_{ij} in the model is the average effective stress of unsaturated soil. This proof method refers to the derivation method in Ref. [55]. Although the idea is relatively concise and some influencing factors (such as body forces, etc.) are ignored, it can reflect the overall proof route and the process of essential development. The concept of π_{ij} as the average effective stress of unsaturated soil will also be reflected in the subsequent discussions.

2.4 Model Summary and Discussion

1. Model Summary

The multi-field coupling model of unsaturated soil proposed in this chapter can be reflected by the average effective stress Eq. (2.70), the shear stress Eq. (2.92), the strain evolution Eq. (2.80), the particle temperature motion Eq. (2.84), and the soil-water characteristic Eq. (2.91), as shown below. According to the given test conditions, the model can simulate and calculate the hydro-thermal coupling characteristics of unsaturated soil. In this model, concepts in classical elastoplastic mechanics such as yield criteria, flow rules, and hardening criteria are not involved. Instead, particle entropy, particle temperature, migration coefficients, and energy functions are used. The model theoretically determines the composition of the dissipative structure of the thermodynamic system and the relationship of migration coefficients, and obtains the constitutive relationship of non-elastic deformation through thermodynamic identities, establishing a relationship between the dissipative mechanism of the soil and the resulting macroscopic physical and mechanical behaviors, as well as the migration coefficient model and the energy function model. These enable the model to more deeply describe the energy dissipation mechanism of unsaturated soil and the resulting macroscopic mechanical characteristics, thus being able to establish a multi-field coupling model of unsaturated soil in complex environments within a unified theoretical framework.

2. Comparison with Other Models

As summarized in the introduction, there are many existing coupling models of soil, and their modeling methods can be roughly classified into three types:

1. Modeling based on classical elastoplastic mechanics theory

Elastoplastic theory is the most widely-applied constitutive theory at present. It divides strain into elastic and plastic parts. Elastic strain is solved according to Hooke's law, and plastic strain is solved according to the plastic increment theory. The Cambridge model is a successful application of elastoplastic theory in soil mechanics. Most of these theories are established on the research basis of metal material tests. Theoretically, only associated flow rules exist. For geotechnical materials, non-associated flow rules should be adopted, but non-associated flow rules do not conform to Drucker's postulate, so there are certain theoretical defects. In addition, the yield functions or hardening parameters involved in the models are usually fitted using simple test results, resulting in a wide variety of constitutive models. Moreover, this type of model may only give well-fitting simulation and prediction results for specimens in a certain characteristic state.

2. Modeling based on superplasticity theory

The superplasticity model defines a dissipation potential function and a free-energy function, and uses the principles of thermodynamics to obtain the flow rules and yield criteria of geotechnical materials. In this way, it can fully satisfy the first and second laws of thermodynamics and avoid the theoretical defects (non-associated flow rules) in the elastoplastic theory model and the arbitrariness of the model function forms. This model is an important supplement to the classical elastoplastic model and provides a more explicit physical meaning for concepts such as flow rules and yield criteria. However, the academic community has different views on the concept of the dissipation potential function. In addition, this theoretical model has not broken away from the construction ideas of the classical elastoplastic mechanics model and still requires the specification of concepts such as yield criteria and flow rules.

3. Modeling based on fluid dynamics theory

The fluid dynamics theory framework is based on physical conservation laws and classical non-equilibrium thermodynamics theory. By constructing thermodynamic identities, combining free-energy functions and energy dissipation laws, the constitutive relations of materials are established, which is widely applied in fluid research. The dissipative system that causes the system to deviate from the equilibrium state can be theoretically determined, and the macroscopic physical and mechanical behaviors can be described through the migration coefficient model (dissipation relation) and the energy function model.

The thermodynamic model of unsaturated soil proposed in this chapter is based on the fluid-dynamics theory framework, more precisely, the fluid-dynamics theory framework of particulate matter. Both the models based on fluid-dynamics theory and those based on superplasticity theory belong to the category of thermodynamic models, but there are the following differences between them: ① Different thermodynamic state variables are adopted. The superplasticity model uses total strain, internal variables, etc., as thermodynamic state variables; the fluid-dynamics model believes that total strain should not be used as a state variable, but elastic strain is used as a state variable, and the evolution of plastic strain is described through dissipation relations. ② The superplasticity model focuses on constructing the dissipation potential function, but does not give the theoretical basis for the selection of the dissipative system; while the fluid-dynamics model can theoretically determine the dissipative force and dissipative flow in the irreversible process to determine the energy-dissipative system. ③ Both models believe that heat generation contains a lot of dissipation-related content, but the specific treatment methods are different. Superplasticity theory usually only gives the heat generation caused by the plastic strain rate and regards the functional relationship between the plastic strain rate and heat generation as another relationship independent of the free-energy function; the fluid-dynamics theory follows the standard treatment method of non-equilibrium thermodynamics, first determining the dissipative system by equilibrium conditions, and then writing heat generation as the product of the dissipative force and the dissipative flow. ④ Although the concept of entropy is also introduced in the superplasticity model, in fact, it does not give a quantitative expression of the evolution of entropy, nor does it consider the changes in variables such as density and temperature in space and time, while these aspects are reflected in the fluid-dynamics model. ⑤ The superplasticity model still needs to specify concepts such as yield criteria and flow rules, while the fluid-dynamics model does not involve these concepts. Its theoretical derivation makes full use of mass, momentum, and energy conservation equations as well as the entropy-increase equation, and is more suitable for establishing a multi-physical-field coupling theory for multi-phase materials.

References

1. Li H, Nagano K, Lai Y (2012) Heat transfer of a horizontal spiral heat exchanger under groundwater advection. Int J Heat Mass Transfer 55(23–24):6819–6831
2. Brandl H (2006) Energy foundations and other thermo-active ground structures. Géotechnique 56(2):81–122
3. Man Y, Yang, Diao N, Liu, Fang Z (2012) A new model and analytical solutions for borehole and pile ground heat exchangers. Int J Heat Mass Transf 53(13–14):2593–2601
4. Uchaipichat A, Khalili N (2009) Experimental investigation of thermo-hydromechanical behaviour of an unsaturated silt. Géotechnique 59(4):339–353
5. Jiang F, Luo L, Chen J (2013) A novel three-dimensional transient model for subsurface heat exchange in enhanced geothermal systems. Int Commun Heat Mass Transf 41:57–62
6. Bellia Z, Ghembaza MS, Belal T (2015) A thermo-hydro-mechanical model of unsaturated soils based on bounding surface plasticity. Comput Geotech 69:58–69
7. Abed AA, Sołowski WT (2017) A study on how to couple thermo-hydro-mechanical behaviour of unsaturated soils: physical equations, numerical implementation and examples. Comput Geotech 92:132–155
8. Rutqvist J, Zheng L, Chen F, Liu HH, Birkholzer J (2014) Modeling of coupled thermo-hydro-mechanical processes with links to geochemistry associated with bentonite backfilled repository tunnels in clay formations. Rock Mech Rock Eng 47(1):167–186

9. Gens A (2010) Soil-environment interactions in geotechnical engineering. Géotechnique 60(1):3–74
10. Alonso EE, Gens A, Josa A (1990) A constitutive model for partially saturated soils. Géotechnique 40(3):405–430
11. Wheeler SJ, Sharma RS, Buisson MSR (2003) Coupling of hydraulic hysteresis and stress-strain behaviour in unsaturated soils. Géotechnique 53(1):41–54
12. Wheeler SJ, Sivakumar V (1995) An elasto-plastic critical state framework for unsaturated soil. Géotechnique 45(1):35–53
13. Sheng DC (2011) Constitutive modelling of unsaturated soils: discussion of fundamental principles. In: Unsatur Soil, vol 1, pp 91–112
14. D'onza F, Gallipoli D, Wheeler S, Casini F, Vaunat J, Khalili N, Laloui L, Mancuso C, Masin D, Nuth M, Pereira JM, Vassallo R (2011) Benchmark of constitutive models for unsaturated soils. Géotechnique 61(4):283–302
15. Khalili N, Loret B (2001) An elasto-plastic model for non-isothermal analysis of flow and deformation in unsaturated porous media: formulation. Int J Solids Struct 38(46–47):8305–8330
16. Yao YP, Hou W, Zhou AN (2009) UH model: three-dimensional unified hardening model for overconsolidated clays. Géotechnique 59(5):451–469
17. Sheng DC, Fredlund DG, Gens A (2008) A new modelling approach for unsaturated soils using independent stress variables. Can Geotech J 45(4):511–534
18. Toll DG (1990) A framework for unsaturated soil behaviour. Géotechnique 40(1):31–44
19. Gallipoli D, Gens A, Sharma R, Vaunat J (2003) An elasto-plastic model for unsaturated soil incorporating the effects of suction and degree of saturation on mechanical behaviour. Géotechnique 53(1):123–135
20. Pereira JM, Wong H, Dubujet P, Dangla P (2005) Adaptation of existing behaviour models to unsaturated states: application to CJS model. Int J Numer Anal Methods Geomech 29(11):1127–1155
21. Wheeler SJ, Sharma RS, Buisson MSR (2003) Coupling of hydraulic hysteresis and stress-strain behaviour in unsaturated soils. Geotechnique 53(1):41–54
22. Tamagnini R (2004) An extended Cam-clay model for unsaturated soils with hydraulic hysteresis. Geotechnique 54(3):223–228
23. Sun DA, Sheng DC, Cui HB, Sloan SW (2007) A density-dependent elastoplastic hydro-mechanical model for unsaturated compacted soils. Int J Numer Anal Methods Geomech 31(11):1257–1279
24. Sheng DC, Zhou AN (2011) Coupling hydraulic with mechanical models for unsaturated soils. Can Geotech J 48(5):826–840
25. François B, Laloui L (2008) ACMEG-TS: a constitutive model for unsaturated soils under non-isothermal conditions. Int J Numer Anal Methods Geomech 32(16):1955–1988
26. Dumont M, Taibi S, Fleureau JM, Abou-Bekr N, Saouab A (2011) A thermo-hydro-mechanical model for unsaturated soils based on the effective stress concept. Int J Numer Anal Methods Geomech 35(12):1299–1317
27. Mašín D, Khalili N (2012) A thermo-mechanical model for variably saturated soils based on hypoplasticity. Int J Numer Anal Methods Geomech 36(12):1461–1485
28. Wu W, Li X, Charlier R, Collin F (2004) A thermo-hydro-mechanical constitutive model and its numerical modelling for unsaturated soils. Comput Geotech 31:155–167
29. Sulem J, Lazar P, Vardoulakis I (2007) Thermo-poro-mechanical properties of clayey gouge and application to rapid fault shearing. Int J Numer Anal Methods Geomech 31(3):523–540
30. Raveendiraraj A (2009) Coupling of mechanical behavior and water retention behavior in unsaturated soils. University of Glasgow, Scotland
31. van Genuchten MT (1980) A closed-form equation for predicting the hydraulic conductivity of unsaturated soils. Soil Sci Soc Am J 44:892–898
32. Wang X, Shao H, Hesser J, Zhang C, Wang W (2014) Numerical analysis of thermal impact on hydro-mechanical properties of clay. J Rock Mech Geotech Eng 6(5):405–416

33. Zhou AN, Sheng DC, Li J (2014) Modelling water retention and volume change behaviours of unsaturated soils in non-isothermal conditions. Comput Geotech 55:1–13
34. Sánchez M, Gens A, Villar MV, Olivella S (2016) Fully coupled thermo-hydro-mechanical double-porosity formulation for unsaturated soils. Int J Geomech 16(6):1–38
35. Houlsby GT, Puzrin AM (2000) A thermomechanical framework for constitutive models for rate-independent dissipative materials. Int J Plast 16(9):1017–1047
36. Li XS (2007) Thermodynamics-based constitutive framework for unsaturated soils_1: theory. Géotechnique 57(5):411–422
37. Li XS (2007) Thermodynamics-based constitutive framework for unsaturated soils_2: a basic triaxial model. Geotechnique 57(5):411–422
38. Muraleetharan K, Liu C, Wei C, Kibbey T, Chen L (2009) An elastoplastic framework for coupling hydraulic and mechanical behavior of unsaturated soils. Int J Plast 25(3):473–490
39. Martin PC, Parodi O, Pershan PS (1972) Unified hydrodynamic theory for crystals, liquid crystals, and normal fluids. Phys Rev A 6:2401–2420
40. Temmen H, Pleiner H, Liu M, Brand HR (2000) Convective nonlinearity in non-Newtonian fluids. Phys Rev Lett 84:3228–3231
41. Jiang YM, Liu M (2009) Granular solid hydrodynamics. Granul Matter 11(3):139–156
42. de Gennes PG (1999) Granular matter: a tentative view. Rev Mod Phys 71(2):374–382
43. Haff PK (1983) Grain flow as a fluid-mechanical phenomenon. J Fluid Mech 134:401–430
44. Jiang YM, Liu M (2010) Hydrodynamic method versus constitutive modeling in geotechnical materials. Rock Soil Mech 31(6):1731–1738
45. Houlsby GT (1981) Study of plasticity theories and their applicability to soils. University of Cambridge, Cambridge
46. Zhang ZC, Cheng XH (2017) A fully coupled THM model based on a non-equilibrium thermodynamic approach and its application. Int J Numer Anal Methods Geomech 41:527–554
47. Zhang ZC, Cheng XH (2016) A thermo-mechanical coupled constitutive model for clay based on extended granular solid hydrodynamics. Comput Geotech 80:373–382
48. Zhang ZC (2017) A thermodynamics-based theory for the thermo-poro-mechanical modeling of saturated clay. Int J Plast 90:164–185
49. Bennethum LS, Cushman JH (1996) Multiscale, hybrid mixture theory for swelling systems-I: balance laws. Int J Eng Sci 34(2):125–145
50. Onsager L (1931) Reciprocal relations in irreversible processes. Phys Rev 37(2):405–426
51. De Groot SR, Mazur P (1984) Non-equilibrium thermodynamics. Dover Publications, New York
52. Jiang YM, Liu M (2014) Granular Solid Hydrodynamics (GSH): a broad-ranged macroscopic theory of granular media. Acta Mech 225:2363–2384
53. Jiang YM, Liu M (2015) Applying GSH to a wide range of experiments in granular media. The European Physical Journal E 38:15
54. Landau LD, Lifshitz EM (1986) Theory of elasticity, 3rd edn. Pergamon Press, Oxford
55. Jiang Y, Einav I, Liu M (2017) A thermodynamic treatment of partially saturated soils revealing the structure of effective stress. J Mech Phys Solids 100:131–146
56. Wong K, Mašín D (2014) Coupled hydro-mechanical model for partially saturated soils predicting small strain stiffness. Comput Geotech 61:355–369
57. Baldi G, Hueckel T, Pellegrini R (1988) Thermal volume changes of the mineral-water system in low-porosity clay soils. Can Geotech J 25(4):807–825
58. Abuel-Naga HM, Bergado DT, Ramana GV, Grino L, Rujivipat P, Thet Y (2006) Experimental evaluation of engineering behavior of soft Bangkok clay under elevated temperature. J Geotech Geoenviron Eng 132:902–910
59. Cekerevac C, Laloui L (2004) Experimental study of thermal effects on the mechanical behaviour of a clay. Int J Numer Anal Methods Geomech 28:209–228
60. Delage P, Sultan N, Cui YJ (2000) On the thermal consolidation of Boom clay. Can Geotech J 37:343–354

61. Towhata I, Kuntiwattanaul P, Seko I, Ohishi K (1993) Volume change of clays induced by heating as observed in consolidation tests. Soils Found 33:170–183
62. Cui YJ, Sultan N, Delage P (2000) A thermomechanical model for saturated clays. Can Geotech J 37(3):607–620
63. Abuel-Naga H, Bergado D, Bouazza A, Ramana G (2007) Volume change behaviour of saturated clays under drained heating conditions: experimental results and constitutive modeling. Can Geotech J 44(8):942–956
64. Tarantino A (2009) A water retention model for deformable soils. Géotechnique 59(9):751–762
65. Lu N, Likos WJ (2004) Unsaturated soil mechanics. Wiley, Hoboken
66. Houlsby GT (1985) The use of a variable shear modulus in elastic-plastic models for clays. Comput Geotech 1(1):3–13
67. Houlsby GT, Amorosi A, Rojas E (2005) Elastic moduli of soils dependent on pressure: a hyperelastic formulation. Geotechique 55(5):383–392
68. Jiang YM, Liu M (2004) Energy instability unjams sand and suspension. Phys Rev Lett 93:148001
69. Jiang YM, Liu M (2007) A brief review of "granular elasticity". Eur Phys J E 22:255–260
70. Lifshitz EM, Landau LD (1987) Fluid mechanics. Pergamon Press, New York
71. Bishop AW, Blight GE (1963) Some aspects of effective stress in saturated and partly saturated soils. Géotechnique 13(3):177–197
72. Houlsby GT (1979) The work input to a granular material. Géotechnique 29(3):354–358
73. Bolzon G, Schrefler BA, Zienkiewicz OC (1996) Elastoplastic soil constitutive laws generalized to partially saturated states. Géotechnique 46(2):279–289
74. Zhao CG, Liu Y, Gao FP (2010) Work and energy equations and the principle of generalized effective stress for unsaturated soils. Int J Numer Anal Methods Geomech 34(9):920–936
75. Gallipoli D (2000) Constitutive and numerical modelling of unsaturated soils. University of Glasgow, Glasgow
76. Kostädt P, Liu M (2000) Causality and stability of the relativistic diffusion equation. Phys Rev D 62(2):1–6
77. Grant S, Salehzadeh A (1996) Calculation of temperature effects on wetting coefficients of porous solids and their capillary pressure functions. Water Resour Res 32(2):261–270
78. Gallipoli D (2012) A hysteretic soil-water retention model accounting for cyclic variations of suction and void ratio. Geotechnique 62(7):605–616
79. Uchaipichat A, Khalili N (2009) Experimental investigation of thermo-hydromechanical behaviour of an unsaturated silt. Geotechnique 59(4):339–353
80. Khalili N, Khabbaz MH (1998) A unique relationship for χ for the determination of shear strength of unsaturated soils. Géotech 48(5):681–688
81. Gallipoli D, Gens A, Chen G, D'onza F (2008) Modelling unsaturated soil behaviour during normal consolidation and at critical state. Comput Geotech 35(6):825–834

Chapter 3
Analysis of Thermo-Hydro-Mechanical Coupling Characteristics of Unsaturated Soil and Model Validation

3.1 Determination of Model Parameters

The unsaturated soil coupling model proposed in this paper involves many thermodynamic concepts and variables, some of which may be uncommon in soil mechanics and cannot be directly measured by soil tests. From the perspective of practical engineering applications, this may lead to problems that are difficult to understand and accept. However, from the perspective of basic scientific research, these quantities objectively exist and are not just empty names and symbols. The parameters involved in the model are as follows: parameters related to the elastic-potential-energy density function, including B_0, c, β_T, ζ, $\lambda(s)$; migration coefficients related to particle motion, including η_v, η_s, λ_{vg}, λ_{sg}, b, γ; and parameters related to the SWCC, a_1, b_1, b_2, n. The specific calibration methods for each parameter are as follows:

1. Parameters related to elastic-potential-energy density

β_T is the soil thermal expansion coefficient, generally taken as 1×10^{-5} °C^{-1}; ζ can be taken as 0.6 with reference to literature [1]; $\lambda(s)$ is the slope of the compression curve under different suctions. According to literature [2], it can be approximately expressed as $\lambda(s) = \lambda(0)[(1 - r)\exp(-\beta s) + r]$. r and β can be determined by fitting the slopes of the compression curves under different suctions. The value of B_0 can be determined by the isotropic compression-rebound curve. At this time, $d_t\varepsilon_s = 0$, and the average net stress can be expressed as:

$$\overline{p} = B_0 \exp\left(-\frac{e}{\lambda(s)}\right)\left(\varepsilon_v^e\right)^{1.5} \tag{3.1}$$

According to the compression-rebound curve, ε_v^e under a certain net stress can be calculated, and then the value of B_0 can be calculated.

G. Yang, *Multi-field Coupled Constitutive Model for Soils Based on Granular Thermodynamics Theory*, https://doi.org/10.1007/978-981-95-0249-3_3

2. Parameters related to particle motion

For the migration coefficients related to particle motion, we do not need to calibrate them one by one, because the parameters required for the final calculation of the model are c_1, c_2, c_3, c_4, $c_5(\lambda_{sg}/\gamma$, $c_3 = \eta_s^{1/\alpha}\lambda_{vg}/\gamma$, $c_4 = \eta_s^{1/\alpha}/b$, $c_5 = \gamma/b)$. Therefore, only the values of $c_1 - c_5$ need to be calibrated. The calibration method for saturated clay in Reference can be referred to [3]. Among them, c_1, c_2, c_3, c_5 can be calibrated under saturated and isothermal conditions, that is, without considering the influence of changes in saturation and temperature.

1. Calibration of c_5

First, calibrate c_5 through a stress-relaxation test. Under isothermal conditions, when a stress-relaxation test is carried out, the strain remains unchanged, $d_t\varepsilon_{ij} = 0$. From Eq. (2.84), we can get $T_{ss} = T_{ss0}\exp(-c_5 t/\rho^S)$, where T_{gg0} is the particle temperature value at the beginning of the relaxation test. After a certain time Δt, the stress relaxation gradually stabilizes, and T_{gg} will gradually approach 0. We can stipulate that when a very small value is reached, for example, $T_{gg} = 10^{-8}T_{gg0}$, the relaxation test ends, and the value of c_5 can be determined accordingly. Since many tests do not involve stress-relaxation tests, we can approximately determine the values of c_5 for different types of soil through the stress-relaxation tests of different types of soil mentioned in Reference [4].

2. Calibration of c_2 and c

For clay, in the critical state, that is, the volume and stress remain unchanged, and the shear strain is still in a continuous development and flowing state. At this time, $d_t\varepsilon_v = 0$, $d_tT_g = 0$, $d_t\varepsilon_{ij}^e = 0$, $\varepsilon_v^e = 0$. Thus, Eq. (2.84) and Eq. (2.81) can be respectively expressed as:

$$d_t\left(T_{ss}\right) = c_5\frac{T_{ss}}{\rho^S} - c_2c_5\frac{d_t\varepsilon_{ij}d_t\varepsilon_{ij}}{\rho^S} \tag{3.2}$$

$$d_t\varepsilon_s^e = d_t\varepsilon_s - \left(T_{gg}\right)^{\alpha}\varepsilon_s^e = 0 \tag{3.3}$$

Subsequently, by combining Eqs. (3.2) and (3.3), we can obtain:

$$\varepsilon_s^e = c_2^{-\alpha}\left(d_t\varepsilon_s\right)^{1-2\alpha} \tag{3.4}$$

As mentioned in Sect. 2.3.5, when α takes the value of 0.5, the elastic shear strain at the critical state is independent of the shear-strain rate, and Eq. (3.4) can be simplified as:

$$\varepsilon_s^e = c_2^{-0.5} \tag{3.5}$$

In addition, at the critical state, $q = Mp'$, where M is the slope of the critical-state line. According to Eqs. (2.70) and (2.92), we know that:

$$\frac{q}{p'} = \frac{\sqrt{6}c^{0.5}\varepsilon_s^e}{0.5c^{-0.5}\left(\varepsilon_s^e\right)^2} = M \tag{3.6}$$

$$q_{cr} = \sqrt{6}B_0 \exp\left(-\frac{v-1}{\lambda(s)}\right)\zeta c^{0.5}\varepsilon_s^e \tag{3.7}$$

c_2 and c can be obtained from Eqs. (3.5), (3.6), and (3.7).

3. Calibration of c_1

When an undrained shear test is carried out, $d_t\varepsilon_v = 0$. At this time, c_1 is the only undetermined parameter required for the calculation. The stress-strain curve and the effective stress path of the undrained shear are controlled by c_1. Based on this, the value of c_1 can be calibrated.

4. Calibration of c_3

According to Eq. (2.84), when an isotropic compression test is carried out, $d_t\varepsilon_s = 0$. At this time, c_3 is the only undetermined parameter, which is related to the change of the volumetric strain. The value of c_3 can be calibrated by the isotropic compression curve.

5. Calibration of c_4

when c_1, c_2, c_3, and c_5 are all determined, c_4 is the only undetermined parameter in the model. For saturated soil, since c_4 is related to the change in temperature, it can be determined by the thermal response of the soil caused by heating, such as the change in thermal volumetric strain. For unsaturated soil, c_4 is also related to the change in saturation. Under isothermal conditions, it can be determined by wetting or drying tests.

3. Parameters related to SWCC

The parameters related to SWCC are a_1, b_1, b_2, n. Among them, $b_2 = -493.5$ °C is a constant, and a_1, b_1, n can be obtained by fitting the test results of SWCC, as shown in Fig. 2.4.

3.2 Analysis of Hydro-Mechanical Coupling Characteristics

In this section, the ability of the proposed model to describe the hydro-mechanical coupling characteristics of unsaturated soil is verified through the test results of two different types of soil, such as the irreversible compression caused by drying during wet-dry cycles and the influence of wet-dry cycles on isotropic consolidation characteristics. The model parameters of the two types of soil are shown in Table 3.1. One is non-swelling Speswhite kaolin [5], and the other is a mixture of highly

Table 3.1 Soil types and model parameters

Thermodynamic parameters					
°C	**Speswhite kaolin**	**Bentonite/kaolin**		**Speswhite kaolin**	**Bentonite/kaolin**
B_0	2.192×10^7 kPa	4.5453×10^6 kPa	c_2	6701.2 s^2	8536.4 s^2
ζ	0.6	0.6	c_3	15,000 s^2	10,000 s^2
c	0.0022	0.0028	c_4	0.150 kg/(m^3·MPa)	0.17 kg/(m^3·MPa)
c_1	0.11	0.05	c_5	37.3 kg/(m^3·s)	20.6 kg/(m^3·s)
SWCC model parameters					
Dryness			**Saturation**		
	Speswhite kaolin	**Bentonite/kaolin**		**Speswhite kaolin**	**Bentonite/kaolin**
a_1	631	1.46	a_1	368	0.716
b_1	0.963	0.243	b_1	0.922	0.175
n	14.8	2.56	n	3.87	2.88

swelling bentonite and kaolin [6]. The SWCC model parameters are taken from Ref. [7].

Since the influence of temperature is not considered, the average net stress and shear stress in the model, as well as the temperature terms in the equations of motion for plastic strain rate and particle temperature can be removed, as shown below:

$$\overline{p} = B\left[\left(\varepsilon_v^e\right)^{1.5} + 0.5\zeta\left(\varepsilon_v^e + c\right)^{-0.5}\left(\varepsilon_s^e\right)^2\right] \tag{3.8}$$

$$q = \sqrt{6}B\zeta\left(\varepsilon_v^e + c\right)^{0.5}\varepsilon_s^e \tag{3.9}$$

$$d_t\varepsilon_{ij}^p = c_1\left(\eta_s^{1/\alpha}T_g\right)^{0.5}\left(\varepsilon_v^e + \frac{ss_r}{K_e}\right)\delta_{ij} + \left(\eta_s^{1/\alpha}T_g\right)^{0.5}e_{ij}^e \tag{3.10}$$

$$d_t\left(T_{gg}\right) = c_2c_5\frac{d_te_{ij}d_te_{ij}}{\rho^S} + c_3c_5\frac{d_t\varepsilon_v d_t\varepsilon_v}{\rho^S} + c_4\frac{sd_ts_r}{\rho^S} - c_5\frac{T_{gg}}{\rho^S} \tag{3.11}$$

Taking the time derivative of Eq. (3.8) and Eq. (3.9), we can obtain:

$$\begin{aligned} d_t\overline{p} = {} & \frac{\rho_L G_s\overline{p}}{\lambda(s)\rho_d}d_t\varepsilon_v + B\left[1.5\left(\varepsilon_v^e\right)^{0.5} - 0.25\zeta\left(\varepsilon_v^e + c\right)^{-1.5}\left(\varepsilon_s^e\right)^2\right]d_t\varepsilon_v^e \\ & + B\zeta\left(\varepsilon_v^e + c\right)^{-0.5}\varepsilon_s^e d_t\varepsilon_s^e \end{aligned} \tag{3.12}$$

$$d_tq = \frac{\rho_L G_s q}{\lambda(s)\rho_d}d_t\varepsilon_v + \frac{\sqrt{6}}{2}B\zeta\left(\varepsilon_v^e + c\right)^{-0.5}\varepsilon_s^e d_t\varepsilon_v^e + \sqrt{6}B\zeta\left(\varepsilon_v^e + c\right)^{0.5}d_t\varepsilon_s^e \tag{3.13}$$

Therefore, based on known experimental conditions, such as the average loading rate, the rate of change of suction, and the loading-unloading path, numerical simulation calculations can be carried out in combination with the above-mentioned equations. In this way, some hydro-mechanical coupling characteristics of unsaturated soil can be simulated.

3.2.1 *Simulation Analysis of the Experiment Conducted by Raveendiraraj [5]*

Raveendiraraj [5] carried out a series of different stress-path tests to study the hydro-mechanical coupling characteristics of unsaturated soil, including wetting and drying tests, as well as isotropic loading-unloading tests. The experimental data of **A9** in the literature are taken to verify the ability of the model proposed in Chap. 2 to describe the hydro-mechanical coupling characteristics of unsaturated soil. The stress paths are shown in Table 3.2. The loading rate of the average net stress in the test is 0.033 kPa/min, the rate of change of suction in the drying and wetting stages is 0.0167 kPa/min, and the slopes of the isotropic compression and rebound curves are $\lambda = 0.177$ and $\kappa = 0.024$, respectively.

Figure 3.1 shows the analysis of the changes in specific volume and degree of saturation with the average net stress during the isotropic loading-unloading test stage. The isotropic loading-unloading process A-B-C is carried out under a constant suction of $s = 300$ kPa, followed by a wet-dry cycle process C-D-E, and then a re-loading process E-F is carried out under the same constant suction of $s = 300$ kPa. It can be seen from the figure that the loading-unloading cycle (A-B-C) leads to irreversible changes in both the soil volume and the degree of saturation. During the wet-dry cycle process, the degree of saturation increases while the specific volume remains almost unchanged (see points C and E). This indicates that there is hysteresis in the soil-water characteristic curve of unsaturated soil, and the change in the degree of saturation during the wet-dry cycle of this test does not cause a change in volume, that is, the yield strength of the unsaturated soil is not reached. In addition, during the subsequent re-loading process, the yield point in the compression curve advances. This is because the wet-dry cycle process increases the degree of saturation of the soil, resulting in a decrease in the yield strength of the soil, further illustrating the influence of the degree of saturation on the mechanical properties of the soil.

Figure 3.2 shows the changes in specific volume and degree of saturation with suction during the wet-dry cycle process under a constant average net stress. After one wet-dry cycle, the specific volume changes little, and the degree of saturation also changes only slightly (about 0.05). It can be seen that, for the simulation results of specific volume changes, the changes in the wetting and drying processes almost

Table 3.2 Stress path for Test A9

Step				
Starting Point	Ending Point	Experimental Content	Average net stress (kPa)	suction (kPa)
A	B	Isotropic loading	10 → 200	300
B	C	Isotropic unloading	200 → 10	300
C	D	Saturation	10	300 → 10
D	E	Dryness	10	10 → 300
E	F	Isotropic loading	10 → 375	300

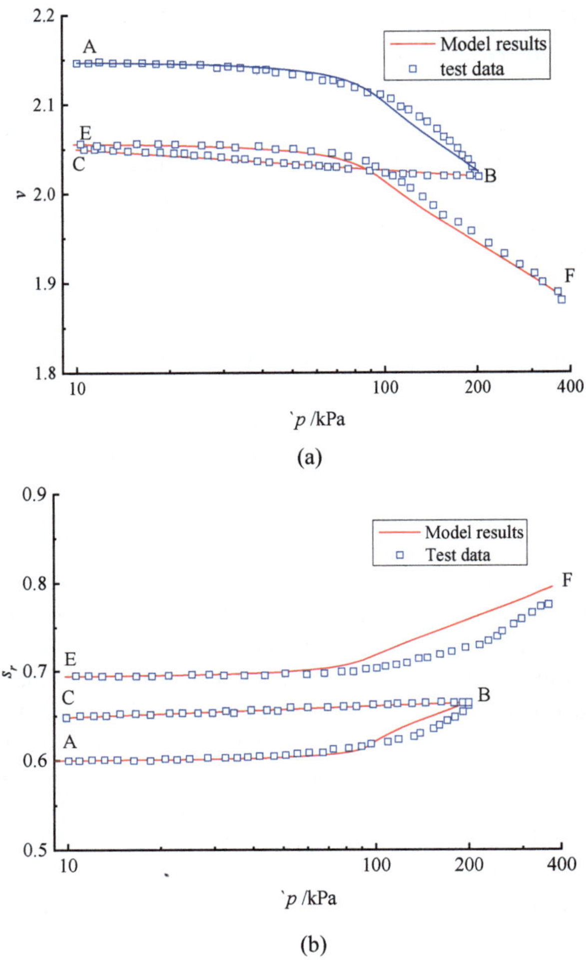

Fig. 3.1 Comparison between predicted results and measured test data for Speswhite kaolin under isotropic loading and unloading: (**a**) specific volume; (**b**) degree of saturation

overlap. For the simulation results of the degree of saturation, the changes in the degree-of-saturation differences can be well reflected, indicating that the model can reflect the changes in the characteristics of unsaturated soil during the wetting-drying process.

In addition, Raveendiraraj [5] also conducted experimental research on the shear characteristics of unsaturated soil. The experimental data of C3 were used to verify

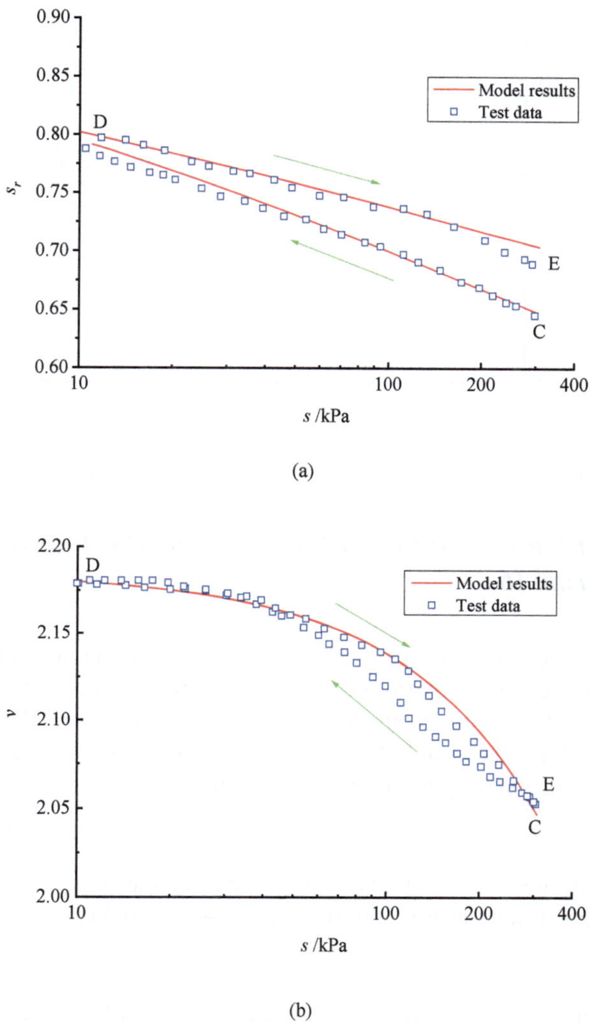

Fig. 3.2 Comparison between predicted results and measured test data for Speswhite kaolin under wetting and drying cycle: (**a**) specific volume; (**b**) saturation degree

the ability of the model to describe the shear characteristics of unsaturated soil, as shown in Fig. 3.3. Drained shear is carried out under a constant axial shear rate of 0.0012 mm/min and an initial confining pressure of 200 kPa. It can be seen that the stress-strain relationship (drainage path $dq = 3dp'$) is basically consistent with the experimental results. Through the comparison between the above-mentioned simulation results and experimental data, it can be seen that the proposed model can well reflect some hydro-mechanical coupling characteristics and shear characteristics of unsaturated soil.

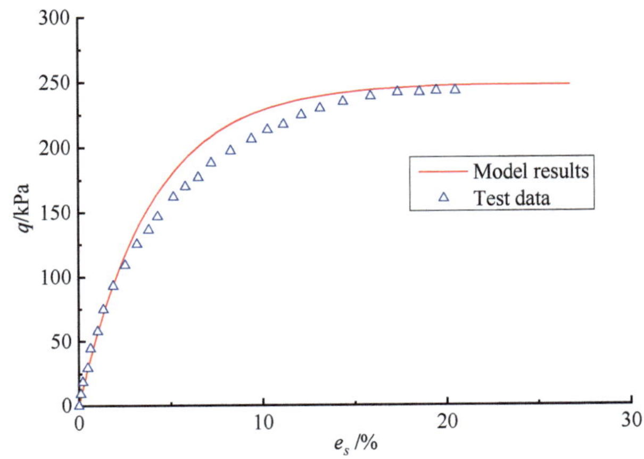

Fig. 3.3 Simulation of stress-strain relationship in triaxial drainage shear test

3.2.2 Simulation Analysis of the Experiment Conducted by Sharma [6]

Sharma [6] carried out a variety of stress-path tests on unsaturated bentonite/kaolinite mixtures to study their hydro-mechanical coupling characteristics. The mixture is composed of 10% bentonite and 90% Speswhite kaolinite, and the model parameters of the soil are listed in Table 3.1. The T8/T9/T10 tests in reference [6] are taken to verify the model, and the stress paths are as follows

T8 test path:

A wetting-drying test is carried out under a constant average net stress of 10 kPa, and the changes in suction are: 300(a) → 20(b) → 300(c), kPa.

T9 test path:

An isotropic loading-unloading test is carried out under a constant suction of 200kpa, and the changes in the average net stress are: 10(a) → 100(b) → 10(c) → 250(d) → 100(e), kPa.

T10 test path:

1. Isotropic loading-unloading is carried out under a constant suction of 200 kPa, and the changes in the average net stress are: 10(a) → 100(b) → 10(c), kPa.
2. A wet-dry cycle test is carried out under a constant average net stress of 10 kPa, and the changes in suction are: 200(c) → 20(d) → 200(e), kPa.
3. An isotropic loading-unloading test is carried out under a constant suction of 200 kPa, and the changes in the average net stress are: 10(e) → 250(f) → 10(g), kPa.

The specific-volume changes in Fig. 3.4a show that there is no collapse phenomenon during wetting (the change from a → b is very small), and irreversible volume shrinkage occurs during the drying process (b → c). This is because the degree of saturation has a large change (about 0.27) during the previous wetting process, as shown in Fig. 3.4b, which increases the effective stress during the drying process and causes the soil to undergo compression deformation.

Figure 3.5 shows the compression-rebound characteristics of the bentonite/kaolinite mixture in the isotropic loading-unloading cycle test. During loading, the degree of saturation increases with the increase of the average net stress, while the change in the degree of saturation during the unloading process is very small, as shown in Fig. 3.5b. In addition, the hysteresis loop of unloading-re-loading indicates the mechanical hysteresis characteristics of unsaturated soil. The yield point

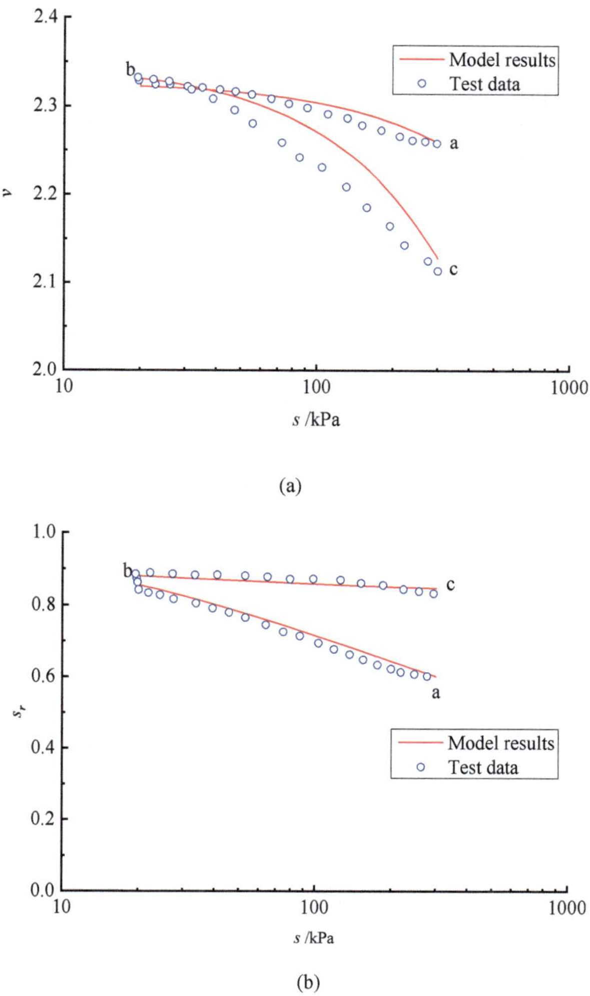

Fig. 3.4 Comparison between predicted results and measured test 8 data for bentonite/kaolin mixture under wetting and drying cycle: (**a**) specific volume; (**b**) saturation degree

occurs at around 100 kPa, corresponding to the maximum average net stress applied previously, as shown in Fig. 3.5a.

Figure 3.6 shows the simulation results of the T10 test. A wet-dry cycle process is inserted between the two isotropic loading processes. The results show that the wet-dry cycle does not cause a change in specific volume, as points c and e in Fig. 3.6a are basically the same. However, during the subsequent re-loading process e → f, the yield point occurs before the previously experienced maximum average net stress, at approximately 80 kPa. Similar to the T9 test, the degree of saturation also increases with the increase of the average net stress, as shown in Fig. 3.6b. However, the values at points c and e are different. After a wet-dry cycle, the degree of saturation increases while the specific volume does not change significantly.

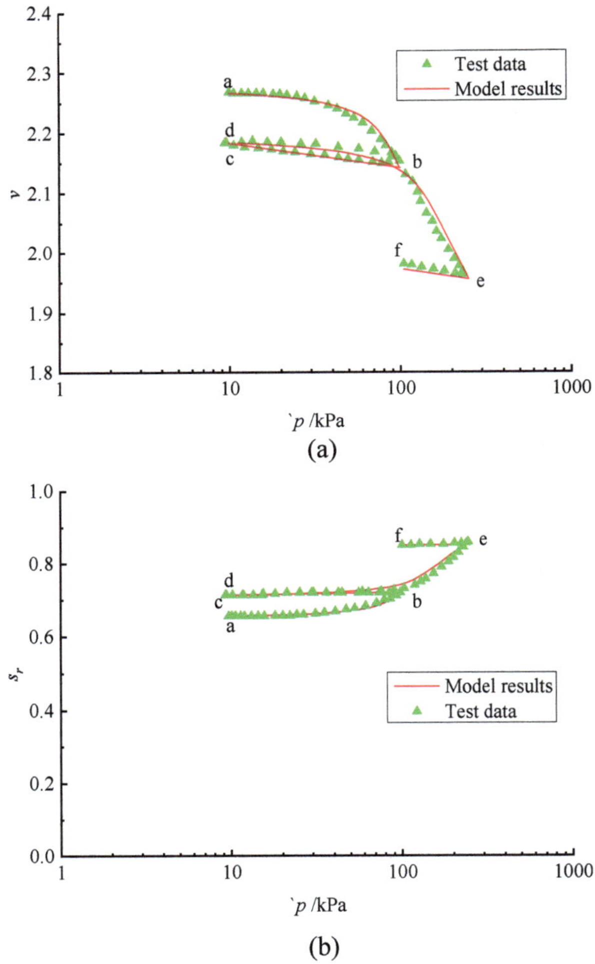

Fig. 3.5 Comparison between predicted results and measured test 9 data for bentonite/kaolin mixture under isotropic loading/unloading cycles: (**a**) specific volume; (**b**) saturation degree

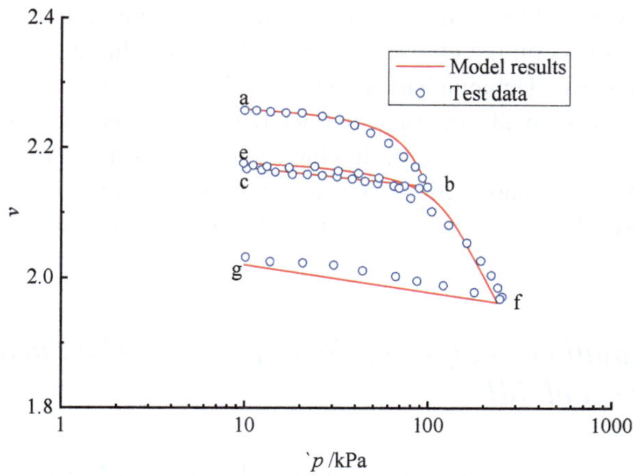

(a)

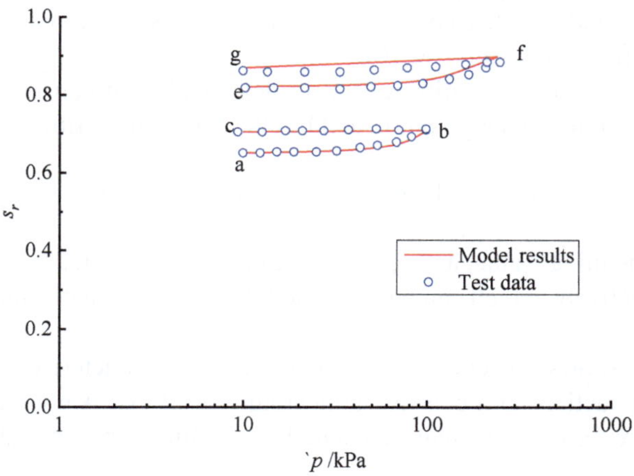

(b)

Fig. 3.6 Comparison between predicted results and measured test 10 data for bentonite/kaolin mixture under isotropic loading/unloading cycles and wetting and drying cycle: (**a**) specific volume; (**b**) saturation degree

3.3 Analysis of Thermo-Hydro-Mechanical Coupling Characteristics

In this section, the ability of the model to describe the thermo-hydro-mechanical coupling characteristics of unsaturated soil is demonstrated through two sets of experimental data. Tang et al. [8] studied the thermo-hydro-mechanical characteristics of unsaturated MX80 bentonite using an isotropic loading device with controllable suction and temperature: the influence of suction and temperature on compressibility was studied through consolidation compression tests; the influence of suction and stress on the characteristics of thermal volume change was studied through heating tests. Uchaipichat and Khalili [9] carried out a series of tests on the thermo-hydro-mechanical coupling characteristics of remolded unsaturated clay during heating, including isotropic loading-unloading tests and shear tests under different temperatures and suctions, water-retention characteristic tests under different temperatures, heating tests with controlled water content, etc.

3.3.1 Simulation Analysis of the Experiment Conducted by Tang et al. [8]

The parameters involved in the model are listed in Table 3.3. Among them, the initial values and the parameters of the slope $\lambda(s)$ of the compression curve are directly taken from Ref. [8], the parameters of the soil-water characteristic curve are taken from Ref. [7], and the thermo-mechanical parameters c_1 – c_5 can be determined according to the tests in Refs. [8, 10].

Since $c_1 = \eta_v/\eta_s$ is a quantity related to deformation, it can be assumed to be affected by suction and temperature, and has the following form:

$$c_1 = c_0\left[1 + c_T\left(T - T_0\right)\right]\exp\left[c_s\left(s - s_0\right)\right] \tag{3.14}$$

where, c_0 is the initial value, and c_T and c_s are parameters related to temperature and suction, respectively, which can be calibrated through tests that control suction and temperature.

Figure 3.7 shows the consolidation-compression characteristics of specimens with different suctions at a constant temperature of 25 °C. Among them, the scattered points are the experimental data, and the solid line is the simulation result. The

Table 3.3 Model parameters

Initial values			
G_S	2.76	S_r	0.5
T_0	25 °C	p_0	0.1 MPa
Thermodynamic model parameters			
c_0	0.001	c_3	9000 s^2
c_4	0.21 kg/(m^3·MPa)	c_5	20.63 kg/(m^3·s)
B_0	8846.48 MPa	$\lambda(0)$	0.25
r	0.18	β	0.14 MPa^{-1}
c_T	0.02 °C^{-1}	c_s	0.005 MPa^{-1}
β_T	1 × 10^{-5} °C^{-1}		
Soil-water characteristic curve parameters			
a_1	0.716	b_1	0.175
n	2.88	b_2	−493.5 °C

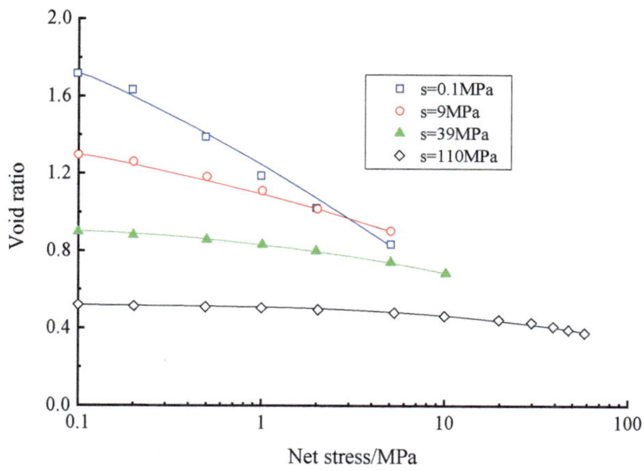

Fig. 3.7 Compression curves at constant suction and at a temperature of 25 °C

suction is kept constant during the test process. It can be seen that as the suction ranges from 0.1 MPa to 110 MPa, the slope of the compression curve decreases. Under high suction, the compressibility is very low, and the slope of the consolidation curve is almost 0. The simulation results are in good agreement with the experimental data.

Figure 3.8 shows the volume changes of specimens with different suctions caused by heating (from 25 °C to 80 °C) under a constant net stress of 0.1 MPa. The test results show that at low suction ($s = 9$ MPa), heating causes thermal contraction; while at high suctions ($s = 39$ MPa, $s = 110$ MPa), heating will cause thermal expansion, and when heated to a certain temperature (about 68 °C when $s = 39$ MPa), it will gradually change from thermal expansion to thermal contraction. For saturated soil, heating the soil sample with a high over-consolidation ratio will produce thermal expansion, while heating the normally consolidated soil and the soil sample with a lower over-consolidation ratio will cause thermal contraction [11–14]. For unsaturated soil in a high-suction state, the soil is in an over-consolidated state, and the higher the suction, the larger the over-consolidation ratio [8]. Therefore, heating unsaturated soil with different suctions will exhibit characteristics similar to those of saturated soil with different over-consolidation ratios. It can be seen that suction has a great influence on the thermal volume change of unsaturated soil, which can be explained by the hardening of unsaturated clay due to the increase in suction.

The thermal volume strain of this model can be given by combining the following formula.

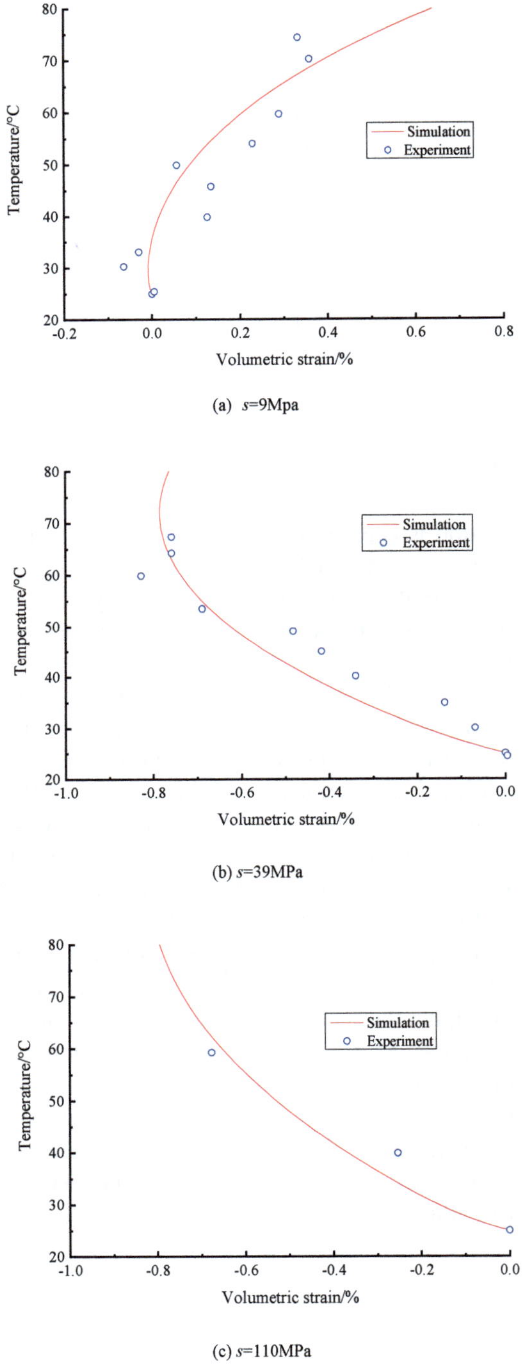

Fig. 3.8 Simulation of volumetric strain changes during heating (from 25 to 80 °C) under constant pressure ($\overline{p} = 0.1$ MPa) for various suction values

$$
\begin{aligned}
\mathrm{d}_t\bar{p} = {} & \frac{\rho_L G_S \bar{p}}{\lambda(s)\rho_d}\mathrm{d}_t\varepsilon_v + B\begin{bmatrix} 1.5\left(\varepsilon_v^e + \beta_T\Delta T\right)^{0.5} \\ -0.25\xi\left(\varepsilon_v^e + \beta_T\Delta T + c\right)^{-1.5}\left(\varepsilon_s^e\right)^2 \end{bmatrix}\mathrm{d}_t\varepsilon_v^e \\
& + B\beta_T\left[1.5\left(\varepsilon_v^e + \beta_T\Delta T\right)^{0.5} - 0.25\xi\left(\varepsilon_v^e + \beta_T\Delta T + c\right)^{-1.5}\left(\varepsilon_s^e\right)^2\right]\mathrm{d}_tT \\
& + B\xi\left(\varepsilon_v^e + \beta_T\Delta T + c\right)^{-0.5}\varepsilon_s^e\mathrm{d}_t\varepsilon_s^e
\end{aligned}
\tag{3.15}
$$

This is the differential form of the average net stress $\bar{p}$. Since $\bar{p}$ remains fixed during heating and there is no shear strain ($\varepsilon_s^e = 0$), Eq. (3.15) can be simplified as:

$$
\frac{\rho_L G_s \bar{p}}{\lambda(s)\rho_d}d_t\varepsilon_v + 1.5B\left(\varepsilon_v^e + \beta_T\Delta T\right)^{0.5}d_t\varepsilon_v^e + 1.5B\beta_T\left(\varepsilon_v^e + \beta_T\Delta T\right)^{0.5}d_tT = 0 \tag{3.16}
$$

It can be seen from Eq. (3.16) that the thermal strain is related to the net stress, the elastic strain and its rate of change, as well as the rate of temperature change. The model proposed in this paper can approximately simulate the experimental results of thermal-strain changes, as shown in Fig. 3.8.

Figure 3.9 shows the comparison of the consolidation-compression curves at different temperatures (25 °C, 80 °C) with the suction controlled constant (s = 9 MPa, 39 MPa, 110 MPa). Among them, the solid lines are the simulation results at 25 °C, and the dashed lines are the simulation results at 80 °C. It can be seen that the slope of the compression curve of unsaturated soil is less affected by temperature and is more determined by the magnitude of suction. Moreover, as the suction increases, compared with the state at 25 °C, the compression curve at 80 °C gradually shifts from below to above, that is, it gradually changes from thermal contraction to thermal expansion, which corresponds to the results in Fig. 3.8. The proposed model can well reflect this change.

3.3.2 Simulation Analysis of the Experiment Conducted by Uchaipichat and Khalili [9]

Uchaipichat and Khalili [9] carried out a series of experimental studies on the THM (thermo-hydro-mechanical) coupling characteristics of unsaturated silty clay. They analyzed the consolidation and shear properties of unsaturated silty clay under different temperatures and suctions. The model parameters involved are listed in Table 3.4.

1. Isotropic Loading-Unloading Tests with Controlled Temperature and Suction

Isotropic loading-unloading tests with constant suction were carried out at three temperatures of 25 °C, 40 °C, and 60 °C, with the suctions controlled at 0, 10, 100, and 300 kPa, respectively. As can be seen from Fig. 3.10, at a certain temperature,

Fig. 3.9 Comparison between two compression curves at constant suction for different temperatures

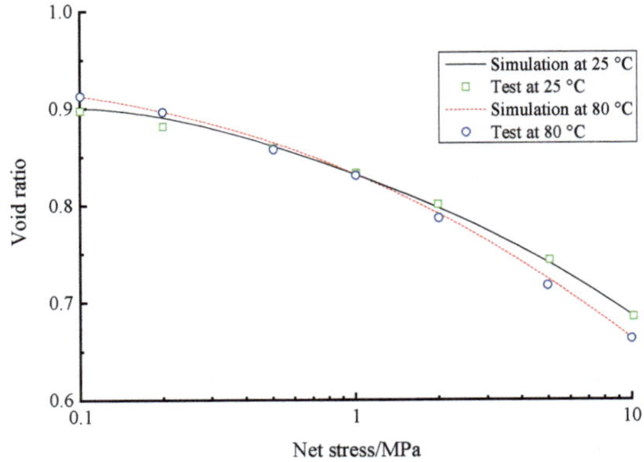

(a) s=9MPa

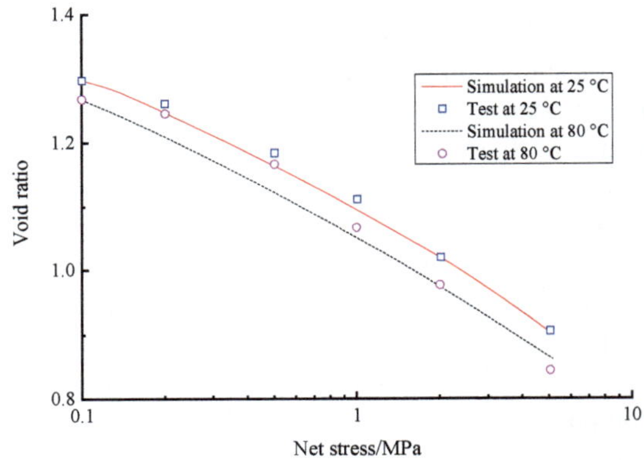

(b) s=39MPa

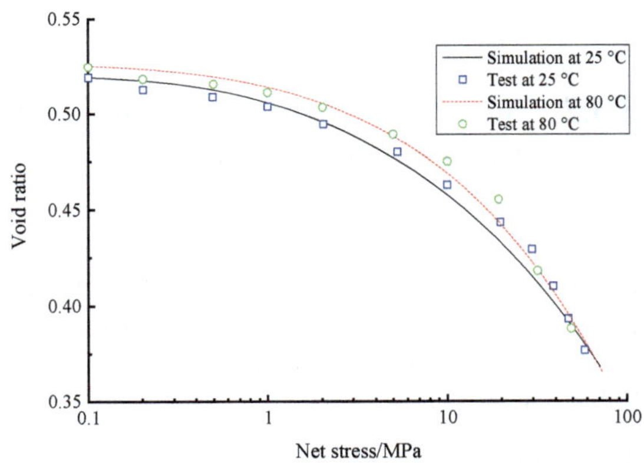

(c) s=110MPa

Table 3.4 Initial values and model parameters

Initial values			
T0	25 °C	p_0	50 kPa
v_0	1.58	S_{r0}	0.48
Basic model parameters			
G_S	2.65	β_T	1×10^{-5} °C^{-1}
c_0	0.29	c_2	7470.36 s^2
c_3	17,000 s^2	c_4	0.25 kg/(m^3·kPa)
c_5	60.34 kg/(m^3·s)	B_0	1.41×10^8 kPa
c	0.0061	M	1.23
c_T	0.0026 °C^{-1}	c_s	0.0022 kPa^{-1}
$\lambda(0)$	0.20		
SWCC parameters			
a_1	9.502	b_1	0.495
n	1.627	b_2	−493.5 °C

as the suction increases, the normal compression curve has a downward-moving trend, and the slopes of the compression curve and the rebound curve decrease with the increase of suction. Moreover, as the temperature increases, this phenomenon becomes more and more obvious. In addition, as the temperature increases, the compression curve also has a downward-moving trend, as shown in Fig. 3.11. The compression curves under different suctions and temperatures are basically consistent. The slopes of the compression and rebound curves decrease slightly with the increase of suction, and basically do not change with the increase of temperature. On the one hand, this may be because when the suction is less than the air-entry value, as the suction increases, the pre-consolidation pressure decreases. On the other hand, the suction hardening phenomenon in the test may also be caused by the structure of different specimens themselves, rather than the influence of the suction itself [9]. The results of the model simulation are in good agreement with the test results, indicating that the proposed model can describe the consolidation characteristics of unsaturated soil under certain suction and temperature conditions.

2. Thermal Volume Strain

The thermal volume strain of the model can be given according to Eq. (3.16). The thermal volume strain is mainly related to the average net stress, the elastic volumetric strain rate, and the heating rate. The simulation results under three different suctions (0, 100, 300 kPa) are shown in Fig. 3.12. For the saturated specimen, heating under a confining pressure of 50 kPa causes thermal expansion, and the volume of the specimen contracts during the subsequent cooling, but a certain amount of irreversible volumetric strain is generated, as shown in Fig. 3.12a. Under a confining pressure of 200 kPa, heating causes thermal contraction and cooling further shrinks the specimen, resulting in a relatively large amount of irreversible deformation, as shown in Fig. 3.12b. This is because the pre-consolidation pressure of all specimens is 200 kPa, which means that the over-consolidation ratio of the specimen under a 50 kPa confining pressure is 4, and the specimen under a 200 kPa confining pressure

Fig. 3.10 Numerical simulations of constant-suction isotropic loading-unloading tests at different temperatures: (**a**) 25 °C; (**b**) 40 °C; (**c**) 60 °C

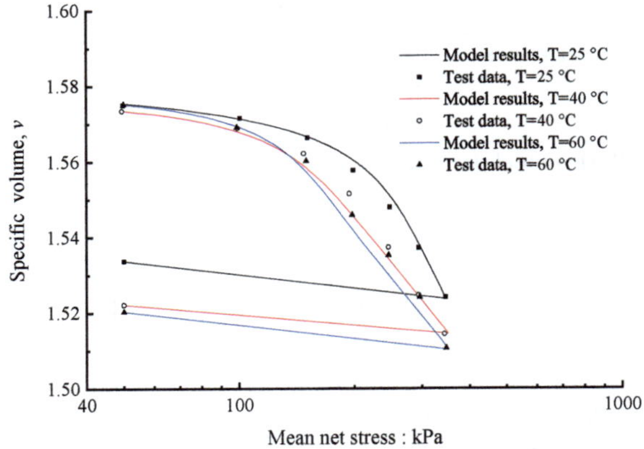

(a)

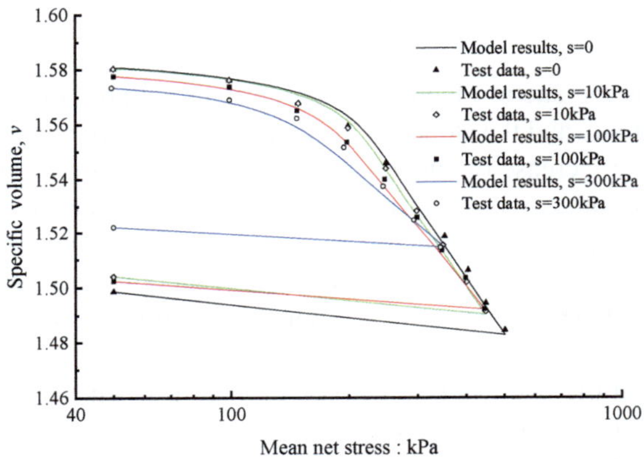

(b)

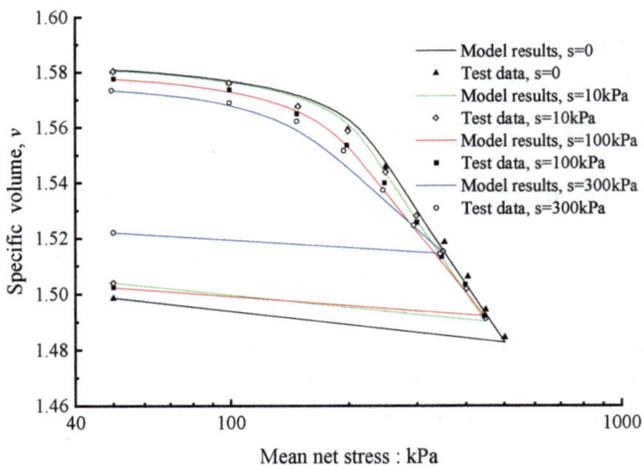

(c)

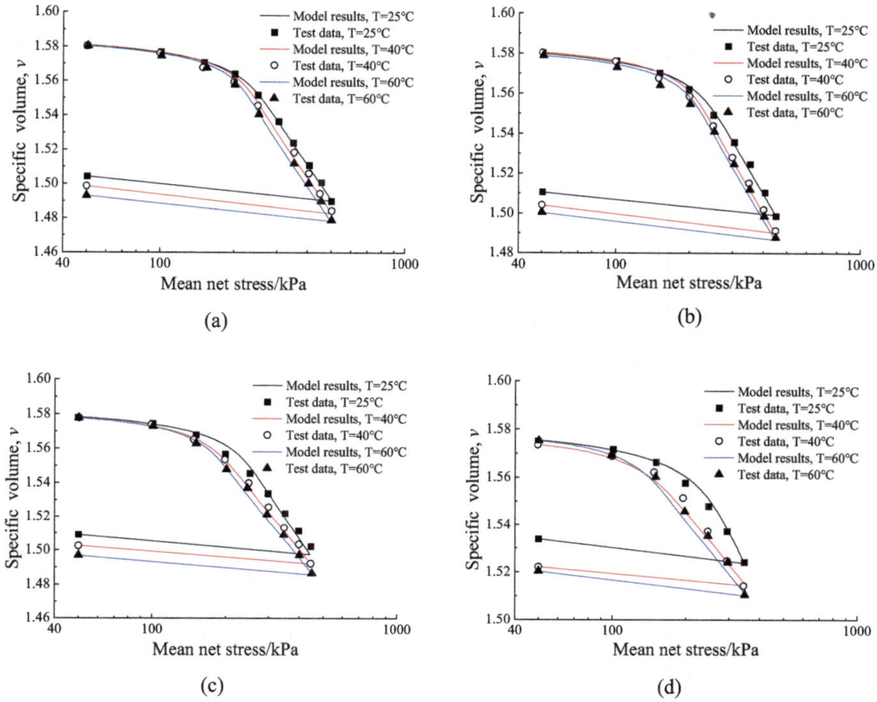

Fig. 3.11 Numerical simulations of isothermal isotropic loading-unloading tests at different matric suctions: (**a**) s = 0; (**b**) s = 10 kPa; (**c**) s = 100 kPa; (**d**) s = 300 kPa

is a normally-consolidated specimen. Studies have shown that for saturated specimens, normally-consolidated specimens and specimens with a low over-consolidation ratio undergo thermal contraction when heated, while specimens with a high over-consolidation ratio undergo thermal expansion when heated [1, 11, 13, 14]. Since the increase in suction also produces a certain over-consolidation effect, similar thermal volume-strain changes are also reflected in unsaturated soil, as shown in Fig. 3.8 in the above text. It can be seen that the simulation results of the volumetric-strain changes caused by the heating-cooling process under different initial confining pressures and suctions are in good agreement with the test results.

3. Shear Characteristics with Controlled Temperature and Suction

The partial-differential form of the shear-stress expression is:

$$\begin{aligned}\mathrm{d}_t q = {} & \frac{\rho_l G_S q}{\lambda(s)\rho_d}\mathrm{d}_t\varepsilon_v + \frac{\sqrt{6}}{2}B\zeta\left(\varepsilon_v^e + \beta_T\Delta T + c\right)^{-0.5}\varepsilon_s^e\mathrm{d}_t\varepsilon_v^e \\ & + \sqrt{6}B\zeta\left(\varepsilon_v^e + \beta_T\Delta T + c\right)^{0.5}\mathrm{d}_t\varepsilon_s^e \\ & + \frac{\sqrt{6}}{2}B\zeta\beta_T\left(\varepsilon_v^e + \beta_T\Delta T + c\right)^{-0.5}\varepsilon_s^e\mathrm{d}_t T\end{aligned} \tag{3.17}$$

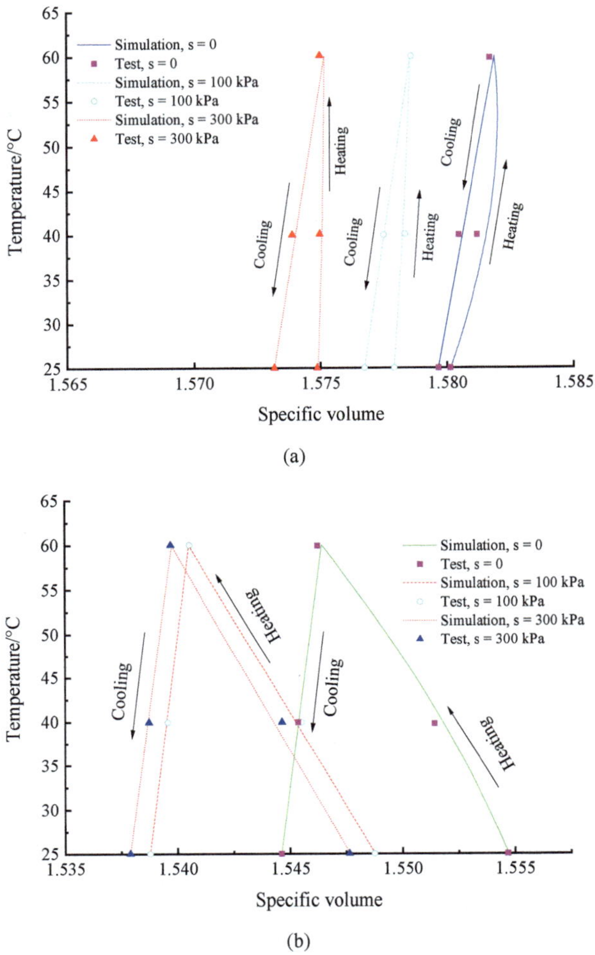

Fig. 3.12 Numerical simulations of specific volume against temperature at different matric suctions with mean net stress: (**a**) 50 kPa; (**b**) 200 kPa

For the triaxial drained shear test, there is the following relationship between the shear stress and the average net stress:

$$d\overline{p} = \frac{1}{3}dq \tag{3.18}$$

According to Eqs. (3.15), (3.17), and (3.18), the evolution expressions of the volumetric strain during the shear process can be obtained. Figure 3.13 shows the variation laws of the shear stress and volumetric strain during the shear process under different temperatures (25 °C, 60 °C) and suctions (0, 100, 300 kPa). It can be seen that as the suction increases, the shear strength of the unsaturated soil also increases, showing hardening; while as the temperature increases, thermal softening

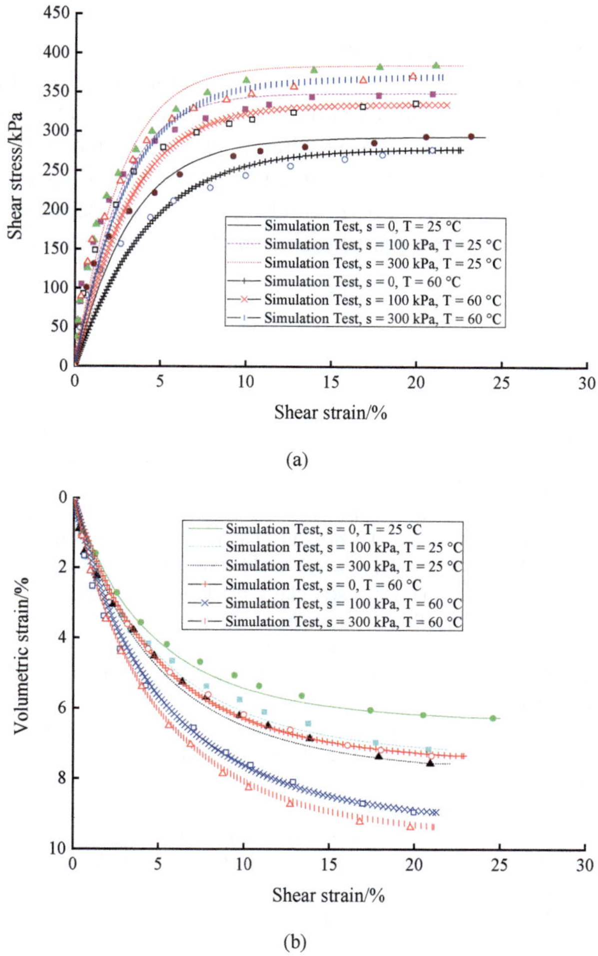

Fig. 3.13 Numerical simulations of suction- and temperature-controlled conventional compression shear tests at initial mean effective stress of 150 kPa: (**a**) deviatoric stress; (**b**) volumetric strain

is exhibited. As the suction and temperature increase, the shear volumetric strain also becomes larger and larger, that is, the contraction phenomenon becomes more and more obvious. From the simulation results, the model can also reflect the shear characteristics of unsaturated soil under such complex stress conditions.

3.4 Summary

This chapter describes in detail the method for determining the parameters in the proposed multi-field coupling constitutive model for unsaturated soil. The model involves many thermomechanical parameters that are not commonly used or accustomed to in soil mechanics. When calculating with the model, there is no need to determine them one by one, and only their combined forms need to be calibrated, and specific calibration methods are provided. By comparing the simulation results of the model with the existing test results, it is proved that the model has the ability to describe the hydraulic and thermo-hydraulic coupling characteristics of unsaturated soil, including the mechanical characteristics after wetting-drying cycles, the consolidation and shear characteristics under different temperatures and suctions, and the thermal volume-strain characteristics under non-isothermal conditions, etc.

References

1. Jiang YM, Liu M (2009) Granular solid hydrodynamics. Granul Matter 11(3):139–156
2. Alonso EE, Gens A, Josa A (1990) A constitutive model for partially saturated soils. Géotechnique 40(3):405–430
3. Yang GC, Bai B (2019) Thermo-Hydro-Mechanical Model for Unsaturated Clay Soils Based on Granular Solid Hydrodynamics Theory. Int J Geomech 19(10): 04019115.
4. Zhu QY, Yin ZY, Zhu JG, Wang JH, Xia XH (2014) Progress and trend of experimental investigation on rate-dependent behavior of soft clays. Rock Soil Mech 5(1): 7–24. (in Chinese)
5. Raveendiraraj A (2009) Coupling of mechanical behaviour and water retention behaviour in unsaturated soils. PhD. Thesis, University of Glasgow, UK
6. Sharma R (1998) Mechanical behavior of unsaturated highly expansive clay. University of Oxford, Oxford, UK
7. Tarantino A (2009) A water retention model for deformable soils. Géotechnique 59(9):751–762
8. Tang AM, Cui YJ, Barnel N (2008) Thermo-mechanical behaviour of a compacted swelling clay. Geotechnique 58(1):45–54
9. Uchaipichat A, Khalili N (2009) Experimental investigation of thermo-hydromechanical behaviour of an unsaturated silt. Géotechnique 59(4):339–353
10. Bishop AW, Blight GE (1963) Some aspects of effective stress in saturated and partly saturated soils. Geotechnique 13(3):177–197
11. Abuel-Naga H, Bergado D, Bouazza A, Ramana G (2007) Volume change behaviour of saturated clays under drained heating conditions: experimental results and constitutive modeling. Can Geotech J 44(8):942–956
12. Sultan N, Delage P, Cui YJ (2002) Temperature effects on the volume change behaviour of Boom clay. Eng Geol 64(2–3):135–145
13. Bai B, Shi XY (2017) Experimental study on the consolidation of saturated silty clay subjected to cyclic thermal loading. Geomech Eng 12(4):707–721
14. Demars KR, Charles RD (1982) Soil volume change induced by temperature cycling. Can Geotech J 19(2):189–194

Chapter 4
Thermo-Hydro-Mechanical Coupling Model of Saturated Clay

4.1 Introduction

As mentioned in the introduction, many practical engineering problems involve the temperature effect of soil, mainly the thermo-hydro-mechanical (THM) coupling problem of rock and soil, which is also a key research problem in the academic field of geotechnical engineering. Currently, many scholars [1–11] have conducted extensive experimental research on this issue and achieved considerable research results. Especially in terms of the thermal response of saturated clay caused by temperature, for example, irreversible thermal strain occurs after a temperature cycle, and the magnitude of thermal strain increases with the number of cycles. Normal consolidated soil will undergo thermal shrinkage when heated, while overconsolidated soil is prone to thermal expansion when heated. As the temperature increases, thermal expansion tends to transform into thermal shrinkage. When the temperature is raised without drainage, it will cause positive pore water pressure, while when the temperature is lowered without drainage, it will produce negative pore water pressure, and the dissipation of pore water pressure is an important factor leading to irreversible volume deformation. In terms of theoretical research, the main focus is on the mechanism of temperature-induced thermal response changes and the study of soil thermal water coupling models. Zhang [12] proposed a THM coupling model for saturated clay based on the theory of particle matter fluid dynamics, which does not require concepts such as flow laws and yield surfaces. Bai Bing et al. [13–15] conducted a systematic study on the thermal solidification characteristics of saturated clay, and obtained the evolution laws of pore water pressure and thermal strain of saturated clay under different temperature paths and boundary conditions. They also derived a theoretical solution method for saturated porous media with hollow cylinders containing internal heat sources.

In engineering construction, due to the complexity of the external environment, the changes in temperature and pressure inside the soil are also very complex.

G. Yang, *Multi-field Coupled Constitutive Model for Soils Based on Granular Thermodynamics Theory*, https://doi.org/10.1007/978-981-95-0249-3_4

Therefore, studying the thermal solidification effect of soil under different temperature paths and confining pressures has great engineering and academic value. In addition, in the second chapter of this article, a theoretical framework for the thermal hydraulic coupling model of unsaturated soil is proposed. In order to verify the ability of the model to describe the thermal solidification characteristics of saturated clay, this chapter is based on the thermal solidification test device developed by Bai Bing et al., and studies the thermal solidification effect of saturated and silty clay samples with different overconsolidation ratios under different confining pressures and temperature paths. The focus is on analyzing the evolution of pore water pressure caused by undrained rise/fall and the strain of drained consolidated bodies over time. Based on the constitutive model framework of thermal water coupling for unsaturated soil proposed in Chap. 2, a thermal water coupling model for saturated soil is proposed for the special case of saturated soil, and the ability of the model to describe the thermal solidification response of saturated silty clay is verified through the obtained experimental results.

4.2 Thermo-Hydro-Mechanical Coupling Model of Saturated Silty Clay

Based on the thermodynamic framework of the multi-field coupling model for unsaturated soil proposed in Chap. 2, this model is degraded to the level of saturated soil for the special case of saturated clay, and a constitutive model that can describe the thermal hydraulic coupling characteristics of saturated clay is established. To verify the ability of the proposed model to describe the thermal solidification characteristics of saturated silty clay considering the overconsolidation effect, this chapter compares the results of thermal solidification tests and cyclic temperature tests with the simulation results of the model.

4.2.1 Basic Assumptions and Thermodynamic State Variables

The basic assumption of saturated soil is similar to that of unsaturated soil: (1) Saturated soil is composed of solid component S (soil particles and their surrounding bound water, which is classified as the solid component) and liquid component L (water in the soil skeleton pores and other liquids). (2) Only when heated, there is a unidirectional transition (solid phase → liquid phase) between the solid and liquid components, and both have the same temperature. (3) Both solid components (soil particles) and liquid components are incompressible, and their intrinsic densities are constant.

Analogous to unsaturated soil, the thermodynamic state variable set of saturated soil is $\left\{\rho^a, m_i^a, S, S_g, \varepsilon_{ij}^e\right\}$, and its corresponding conjugate is $\left\{\mu_{ca}, v_i^a, T, T_g, \pi_{ij}\right\}$, $a \in [S, L]$.

4.2.2 Basic Equation

1. Mass-conservation equation

Without considering the external mass supply of solid and liquid phases, and there is a unidirectional (solid liquid) mass exchange $\overline{\rho}$ between the solid and liquid phases due to temperature changes. Compared with soil particles, the proportion of bound water is very small, and the bulk density of the solid phase containing bound water is approximately taken as $\rho^S = n^S \rho_s$. Therefore, the mass conservation equations for the solid and liquid phases of saturated soil can be expressed as:

$$d_t \rho^S = \rho^S \psi_{kk}^s - \overline{\rho} \tag{4.1}$$

$$d_t \rho^L = \rho^L \psi_{kk}^L - v_i^{LS} \nabla \rho^L + \overline{\rho} \tag{4.2}$$

2. Momentum conservation equation

On the basis of Eq. (2.25), removing the gas phase component, the momentum conservation equation of saturated soil can be expressed as:

$$\sum_{a=S,L} \left(\rho^a d_t v_i^a + \nabla_j n^a \sigma_{ij}^a - \rho^a g\right) + \rho^L v_k^{LS} \nabla_k v_i^L = 0 \tag{4.3}$$

3. Energy conservation equation

Similarly, based on Eq. (2.28), the energy conservation equation for saturated soil can be expressed by removing the gas phase component:

$$\sum_{a=S,L} \left(\rho^a d_t e_a\right) = \sum_{a=S,L} \left(n^a \sigma_{kj}^a \psi_{jk}^a\right) + \nabla_k Q_k - \rho^L v_k^{LS} \nabla_k e_L \tag{4.4}$$

4. Entropy increase equation

Similarly, the equations of motion for thermodynamic entropy and particle entropy of saturated soil are:

$$\sum_{a=S,L} \left(\rho^a d_t \vartheta_a\right) = R / T + \nabla_k f_k - \rho^L v_k^{LS} \nabla_k \vartheta_L \tag{4.5}$$

$$\rho^S d_t \vartheta_g = R_g / T_g - I_g \tag{4.6}$$

4.2.3 Determination of Dissipative System

Similar to unsaturated soil, the set of dissipative forces in saturated soil is $\{\pi_{ij}, T_g, \nabla_i T, d_t\varepsilon_{ij}, \psi_{ij}^L\}$, and the corresponding dissipative flow is $\{Y_{ij}, I_g, q_k / T, \sigma_{ij}^{vS}, \sigma_{ij}^{vL}\}$.

Then, the Onsager form of the entropy production rate R can be expressed as:

$$R = Y_{ij}\pi_{ij} + I_g T_g + \frac{q_k}{T}\nabla_i T + \sigma_{ij}^{vS} d_t\varepsilon_{ij} + \sigma_{ij}^{vL}\psi_{ij}^L \tag{4.7}$$

Similarly, considering the viscosity and temperature effects at the particle level, the particle entropy yield can be expressed as:

$$R_g = \sigma_{ij}^{vg} d_t\varepsilon_{ij} + \Re \mathrm{T}_g d_t T \tag{4.8}$$

where, $RT_g d_t T$ represents the dissipation at the particle level caused by the influence of temperature on saturated soil, $d_t T$ is the dissipation force, RT_g represents dissipative flow. When $T_g = 0$ occurs, the dissipation also disappears.

Similarly, dissipative flow can be expressed as a linear function of dissipative force, in the following form:

$$\begin{Bmatrix} Y_{ij} \\ I_g \\ q_i \\ \sigma_{ij}^{vS} \\ \sigma_{ij}^{vL} \end{Bmatrix} = \begin{bmatrix} \eta_{ijkl} & 0 & 0 & 0 & 0 \\ 0 & \gamma & 0 & 0 & 0 \\ 0 & 0 & \kappa_{ik} & 0 & 0 \\ 0 & 0 & 0 & \lambda_{ijkl}^S & 0 \\ 0 & 0 & 0 & 0 & \lambda_{ijkl}^L \end{bmatrix} \begin{Bmatrix} \pi_{kl} \\ T_g \\ \nabla_k T \\ d_t\varepsilon_{kl} \\ \psi_{kl}^L \end{Bmatrix} \tag{4.9}$$

Similarly, dissipation at the particle level can also be expressed as:

$$\sigma_{ij}^{vg} = \lambda_{ijkl}^g d_t\varepsilon_{kl} \tag{4.10}$$

4.2.4 Thermodynamic Identity

The full differential thermodynamics of saturated soil can be expressed in the form of specific energy e as:

$$\sum_{a=S,L}\left(\rho^a d_t e_a\right) = \pi_{ij} d_t\varepsilon_{ij}^e + T_g \rho^S d_t\vartheta_g + \sum_{a=S,L}\left(\rho^a \mu_a d_t\rho^a + T\rho^a d_t\vartheta_a\right) \tag{4.11}$$

where, $\mu_a = \left(\mu_{ca} + T\vartheta_a + T_g\vartheta_a + v_i^a v_i^a / 2 - e_a\right) / \rho^a$, only when $a = S$, the $T_g\vartheta_g$ term is included. For the liquid phase, the fluid mechanics pressure

$P_L = (\rho^L)^2 \mu_L = \rho^L \mu_{cL} + v_i^L m_i^L + Tn^L s_L - n^L w_L$ is the pore water pressure u_l. For solid phase, $P_S = (\rho^S)^2 \mu_S = \rho^S \mu_{cS} + v_i^S m_i^S + Tn^S s_S - n^S w_S$, P_S represent the response of soil particles to isotropic compression by pore water.

By substituting the conservation equation and entropy increase equation into Eq. (4.11), we can obtain:

$$\begin{aligned} R + R_g &= \left[n^S \sigma_{ij}^S - \pi_{ij} - \mu_S (\rho^S)^2 \delta_{ij} \right] \psi_{ij}^S \\ &+ \left[n^L \sigma_{ij}^L - \mu_L (\rho^L)^2 \delta_{ij} \right] \psi_{ij}^L + \nabla_k (Q_k - f_k T) \\ &+ \left(\rho^L \mu_L \nabla_k \rho^L + T \rho^L \nabla_k \vartheta_L - \rho^L \nabla_k e_L \right) v_k^{LS} \\ &+ \pi_{ij} d_t \varepsilon_{ij}^p + T_g I_g + f_k \nabla_k T + \left(\rho^S \mu_S - \rho^L \mu_L \right) \hat{\rho} \end{aligned} \tag{4.12}$$

The entropy generation rate can also be expressed in the form of Eq. (4.7) plus Eq. (4.8):

$$R + R_g = Y_{ij} \pi_{ij} + I_g T_g + \frac{q_k}{T} \nabla_i T + \left(\sigma_{ij}^{vS} + \sigma_{ij}^{vg} \right) \psi_{ij}^S + \sigma_{ij}^{vL} \psi_{ij}^L + \Re T_g d_t T \tag{4.13}$$

Similarly, by comparing Eqs. (4.12) and (4.13), we can obtain the thermodynamic relationship equation:

$$\Re T_g d_t T = \left(\rho^S \mu_S - \rho^L \mu_L \right) \bar{\rho} \tag{4.14}$$

where, $\bar{\rho} = \rho^L \alpha_{LT} d_t T$, α_{LT} is the rate at which bound water is converted into free pore water during heating, which can be taken as 0.002 °C^{-1}. Substituting it into Eq. (4.14) yields:

$$\Re T_g = \left(\rho^S \mu_S - \rho^L \mu_L \right) \rho^L \alpha_{LT} = \left(P_S / \rho^S - u_l / \rho^L \right) \rho^L \alpha_{LT} \approx \alpha_{LT} \alpha_v p' \tag{4.15}$$

where, p' represents the average effective stress; α_v is the proportionality coefficient.

By comparing with Sect. 2.3.3.3 of Chap. 2, it can be seen that Eq. (4.15) has the same expression form as Eq. (2.61), which can also be seen as the degradation of the unsaturated soil model to the saturated soil model form.

4.2.5 Determination of Stress Expression

On the basis of Eq. (2.64), the elastic potential energy density function of saturated clay can be taken as:

$$w_e = B \left[\frac{2}{5} \left(\varepsilon_v^e + \beta_T \Delta T \right)^{2.5} + \zeta \left(\varepsilon_v^e + \beta_T \Delta T + c \right)^{0.5} \left(\varepsilon_s^e \right)^2 \right] \tag{4.16}$$

where, $B = B_0 \exp\left(-\frac{\nu-1}{\lambda}\right)$, λ is the slope of the saturated soil compression curve.

Thus, the expression for π_{ij} can be derived:

$$\pi_{ij} = \frac{\partial w_e}{\partial \varepsilon_{ij}^e} = B\begin{bmatrix} \left(\varepsilon_v^e + \beta_T \Delta T\right)^{1.5}\delta_{ij} + 0.5\zeta\left(\varepsilon_v^e + \beta_T \Delta T + c\right)^{-0.5}\left(\varepsilon_s^e\right)^2 \delta_{ij} \\ +2\zeta\left(\varepsilon_v^e + \beta_T \Delta T + c\right)^{0.5} e_{ij}^e \end{bmatrix} \tag{4.17}$$

Similarly, let the elastic bulk modulus K_e and elastic shear modulus G_e be defined as:

$$K_e = B\left[\left(\varepsilon_v^e + \beta_T \Delta T\right)^{0.5} + 0.5\zeta\left(\varepsilon_v^e + \beta_T \Delta T + c\right)^{-0.5}\frac{\left(\varepsilon_s^e\right)^2}{\varepsilon_v^e + \beta_T \Delta T}\right] \tag{4.18}$$

$$G_e = B\zeta\left(\varepsilon_v^e + \beta_T \Delta T + c\right)^{0.5} \tag{4.19}$$

Therefore, π_{ij} can also be expressed as:

$$\pi_{ij} = K_e\left(\varepsilon_v^e + \beta_T \Delta T\right)\delta_{ij} + 2G_e e_{ij}^e \tag{4.20}$$

The average effective stress p' and shear stress q can be expressed as:

$$p' = \frac{\pi_{kk}}{3} = B\left[\left(\varepsilon_v^e + \beta_T \Delta T\right)^{1.5} + 0.5\zeta\left(\varepsilon_v^e + \beta_T \Delta T + c\right)^{-0.5}\left(\varepsilon_s^e\right)^2\right] \tag{4.21}$$

$$q = \sqrt{\frac{3}{2}s_{ij}s_{ji}} = \sqrt{6}B\zeta\left(\varepsilon_v^e + \beta_T \Delta T + c\right)^{0.5}\varepsilon_s^e \tag{4.22}$$

4.2.6 Migration Coefficient Relationship

The transfer coefficients η_{ijkl} and λ_{ijkl}^g are defined as:

$$\eta_{ijkl} = \left(T_g\right)^\alpha \frac{\eta_s}{2G_e}\delta_{il}\delta_{jk} + \left(T_g\right)^\alpha\left(\frac{\eta_v}{3K_e} - \frac{\eta_s}{6G_e}\right)\delta_{ij}\delta_{lk} \tag{4.23}$$

$$\lambda_{ijkl}^g = \lambda_{sg}T_g\delta_{il}\delta_{jk} + \left(\lambda_{vg} - \frac{\lambda_{sg}}{3}\right)T_g\delta_{ij}\delta_{lk} \tag{4.24}$$

where, α is also taken as 0.5.

Therefore, the plastic strain rate can be expressed as:

$$d_t\varepsilon_{ij}^{p} = \eta_{ijkl}\pi_{kl} = c_1\left(\eta_s^{1/\alpha}T_g\right)^{\alpha}\left(\varepsilon_v^{e} + \beta_T\Delta T\right)\delta_{ij} + \left(\eta_s^{1/\alpha}T_g\right)^{\alpha} e_{ij}^{e} \tag{4.25}$$

where, $c_1 = \eta_v/\eta_s$.

Similarly, the equation of motion for particle temperature can also be derived:

$$d_t\left(T_{gg}\right) = c_2c_5\frac{d_te_{ij}d_te_{ij}}{\rho^S} + c_3c_5\frac{d_t\varepsilon_v d_t\varepsilon_v}{\rho^S} + c_4\frac{\alpha_{LT}p'd_tT}{\rho^S} - c_5\frac{T_{gg}}{\rho^S} \tag{4.26}$$

where, $c_2 = \eta_s^{1/\alpha}\lambda_{sg}/\gamma$, $c_3 = \eta_s^{1/\alpha}\lambda_{vg}/\gamma$, $c_4 = \alpha_v\eta_s^{1/\alpha}/b$, $c_5 = \gamma/b$, $T_{gg} = \eta_s^{1/\alpha}T_g$.

4.3 Simulation Analysis of Experimental Results

The thermal hydraulic coupling constitutive model for saturated soil presented in this chapter is a degradation of the multi-field coupling constitutive model for unsaturated soil proposed in Chap. 2. Therefore, the method for determining the model parameters involved is similar to that in the previous chapter, as detailed in Sect. 3.1. The specific model parameters are listed in Table 4.1. Since this experiment does not involve shear characteristics, data and parameters related to shear tests can be found in reference [16]. The thermal response test results of saturated silty clay with different overconsolidation ratios obtained in this chapter, as well as some previous thermal consolidation test results conducted by our research group, are used to verify the ability of the proposed model to describe the thermal consolidation characteristics of saturated soil.

4.3.1 Thermal Consolidation Test Considering Overconsolidation Ratio

1. Simulation analysis of thermal strain

Table 4.1 Model parameters

Initial value			
G_S	2.72	T_0	25 °C
p_0	50 kPa	M	0.8
λ	0.21	β_{T0}	1×10^{-5} °C^{-1}
θ	1×10^{-7} °C^{-2}		
Thermodynamics parameters			
c_0	0.036	c_T	0.005 °C^{-1}
c_3	9000 s^2	c_4	0.003 kg/(m^3·kPa)
c_5	50.56 kg/(m^3·s)	c	0.001
B_0	1.47×10^5 kPa		

The volumetric strain given in Sect. 4.2 of the experiment is calculated based on the drainage volume of the specimen. In fact, the true volumetric strain of the specimen is equal to the drainage strain minus the corresponding thermal expansion strain of pore water and soil particles, that is:

$$\varepsilon_v = \frac{\Delta V_{dr} - \left(\alpha_w V_w + \alpha_s V_s\right)\Delta T}{V} \tag{4.27}$$

where, α_w and α_s are the thermal expansion coefficients of pore water and soil particles, respectively, which increase with temperature. The thermal expansion coefficient of pore water is about 17 times larger than that of soil particles, and the average value is taken $\alpha_w = 4.76 \times 10^{-4}$ °C^{-1}, $\alpha_s = 2.80 \times 10^{-5}$ °C^{-1}; V_w and V_s are the volumes of pore water and soil skeleton, respectively; V_{dr} is the total drainage volume; V is the total volume of the saturated sample.

In Sect. 4.2, the volumetric strain given in the experiment needs to be corrected according to Eq. (4.27), and the corrected experimental results are used for comparison and analysis with the simulation calculation results.

The effective stress principle of Taisha Ji can be expressed as:

$$\sigma = p' + u_l \tag{4.28}$$

In the hot bonding test, the confining pressure remains constant, that is $d_t\sigma = 0$, it can be obtained:

$$d_t u_l = -d_t p' \tag{4.29}$$

According to Eq. (4.21), the differential form of effective stress is:

$$\begin{aligned} d_t p' = {} & \frac{\rho_L G_S p'}{\lambda \rho_d} d_t \varepsilon_v + B\left[1.5\left(\varepsilon_v^e \beta_T \Delta T\right)^{0.5} - 0.25\zeta\left(\varepsilon_v^e \beta_T \Delta T + c\right)^{-1.5}\left(\varepsilon_s^e\right)^2\right] d_t \varepsilon_v^e \\ & + B\beta_T\left[1.5\left(\varepsilon_v^e + \beta_T \Delta T\right)^{0.5} - 0.25\zeta\left(\varepsilon_v^e \beta_T \Delta T + c\right)^{-1.5}\left(\varepsilon_s^e\right)^2\right] d_t T \\ & + B\zeta\left(\varepsilon_v^e \beta_T \Delta T + c\right)^{-0.5} \varepsilon_s^e d_t \varepsilon_s^e \end{aligned} \tag{4.30}$$

The experiment does not consider shear action, and Eq. (4.30) can be simplified as:

$$d_t p' = \frac{\rho_L G_S p'}{\lambda \rho_d} d_t \varepsilon_v + 1.5B\left(\varepsilon_v^e + \beta_T \Delta T\right)^{0.5} d_t \varepsilon_v^e + 1.5B\beta_T\left(\varepsilon_v^e + \beta_T \Delta T\right)^{0.5} d_t T \tag{4.31}$$

In the process of drainage consolidation, it can be considered that the pore pressure is constant, that is $dp' = 0$, the expression for the rate of change of volumetric strain can be obtained using Eq. (4.31):

$$\frac{\rho_L G_S p'}{\lambda \rho_d} d_t \varepsilon_v + 1.5B\left(\varepsilon_v^e + \beta_T \Delta T\right)^{0.5} d_t \varepsilon_v^e + 1.5B\beta_T\left(\varepsilon_v^e + \beta_T \Delta T\right)^{0.5} d_t T = 0 \tag{4.32}$$

From Eq. (3.14), it can be inferred that parameter c_1 is related to temperature changes and can be assumed to have a linear relationship:

$$c_1 = c_0\left[1 + c_T\left(T - T_0\right)\right] \tag{4.33}$$

The denser the soil, the greater its thermal expansion coefficient, so β_T will increase with the increase of OCR. In addition, research has shown that β_T also varies with temperature, and the expression for β_T related to OCR and temperature given by Laloui and Cekerevac [17] is:

$$\beta_T = \left(\beta_{T0} + \theta \Delta T\right) \cdot OCR \tag{4.34}$$

where, is the thermal expansion coefficient of normally consolidated soil at room temperature, which can be taken as 1×10^{-5}°C^{-1}; θ is a temperature related parameter.

Figure 4.1 shows the simulation results of the consolidation compression rebound test of normally consolidated saturated silty clay at room temperature. Due to the fact that the experiment first conducted undrained heating and then drained consolidation after the pore pressure stabilized, the initial strain of the drained body rapidly increased and then stabilized. When calculating the model, the final change in volumetric strain is taken without considering the factor of time. It can be seen that the simulation results are consistent with the experimental data, and the slope of the rebound curve is basically the same. By unloading rebound, samples with different overconsolidation ratios can also be obtained, such as samples with overconsolidation ratios of 4 and 8 obtained in this experiment.

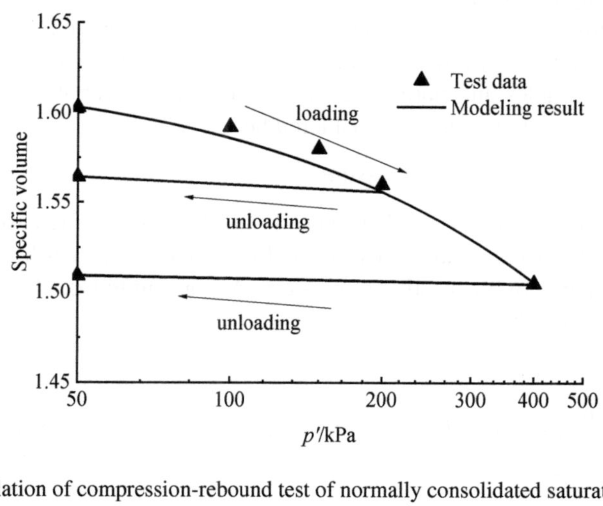

Fig. 4.1 Simulation of compression-rebound test of normally consolidated saturated soil sample

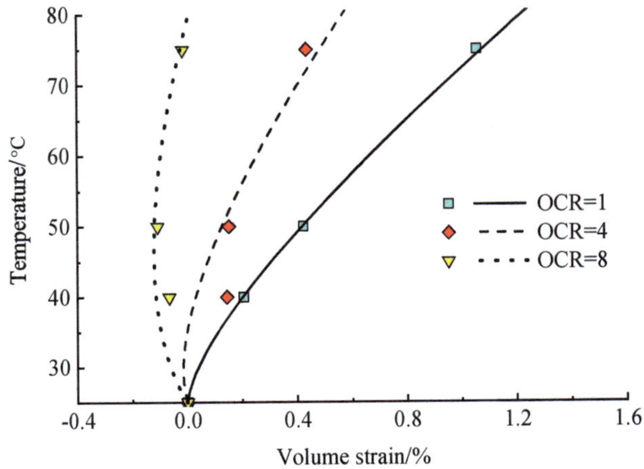

Fig. 4.2 Simulation of volume strain change of samples with different overconsolidation ratios during heating

Figure 4.2 shows the simulation results of the variation of thermal strain with temperature for specimens with different overconsolidation ratios during heating. For normally consolidated soil, heating will cause thermal shrinkage, while as the overconsolidation ratio increases, heating will cause thermal expansion. In addition, for specimens with a high degree of overconsolidation, when heated to a certain temperature, thermal expansion will gradually transition to thermal shrinkage, and many scholars have obtained similar experimental phenomena [5, 6, 13, 18]. As mentioned earlier, the thermal expansion coefficient β_T is relatively large for specimens with high overconsolidation ratios. Based on this, the proposed model can be used to simulate the changes in thermal strain of specimens with different overconsolidation ratios when heated. As shown in Fig. 4.2, the simulation results can reflect the changing trend of thermal strain. For specimens with different overconsolidation ratios, cooling can be regarded as an elastic change, independent of stress history and applied confining pressure, but related to the thermal expansion coefficients of soil particles and pore water [6]. Therefore, the variation of body strain during the cooling process is not considered here.

2. Simulation analysis of pore pressure

Under isobaric conditions, when there is no drainage to increase or decrease temperature, excess pore water pressure will be generated. According to the effective stress principle of Taisha Ji, $d_t u_l = -d_t p'$. According to Eq. (4.31), the expression for $d_t u_l$ can be obtained:

$$d_t u_l = -\frac{\rho_L G_S p'}{\lambda \rho_d} d_t \varepsilon_v - 1.5B\left(\varepsilon_v^e + \beta_T \Delta T\right)^{0.5} d_t \varepsilon_v^e - 1.5B\beta_T\left(\varepsilon_v^e + \beta_T \Delta T\right)^{0.5} d_t T \quad (4.35)$$

Due to the lack of drainage in the experiment, $d_t\varepsilon_v = 0$, Eq. (4.35) can be simplified as:

$$d_t u_l = 4.5Bc_1\left(\varepsilon_v^e + \beta_T\Delta T\right)^{1.5}\left(T_{gg}\right)^a - 1.5B\beta_T\left(\varepsilon_v^e + \beta_T\Delta T\right)^{0.5} d_t T \quad (4.36)$$

Figure 4.3 shows the simulation of the variation of pore pressure over time without drainage heating. The heating rate of the experiment is 0.5 °C/min, based on which the variation of pore pressure with time can be calculated when the temperature is not drained. From the experimental results, it can be seen that as the overconsolidation ratio increases, the maximum pore pressure reached by heating tends to decrease, and the rate of increase in pore pressure also tends to decrease. The simulation results can reflect this variation pattern of pore pressure. Figure 4.4 shows the simulated changes in pore pressure when the temperature rises to 50 °C and 75 °C. It can be seen that the simulation results at a temperature of 50 °C are larger than the experimental values, and the simulation results at 75 °C are closer to the experimental values, which is related to the higher pore pressure growth rate in the simulation results shown in Fig. 4.3.

4.3.2 Thermal Response Analysis Generated by Cyclic Temperature Load

Our research group conducted experimental studies on the consolidation characteristics of saturated silty clay under repeated hot and cold cycling temperatures [13], with a focus on analyzing the effects of moisture content, confining pressure, thermal load size, and cycling times. The experimental results indicate that under undrained conditions, as the moisture content and thermal load increase, the pore water

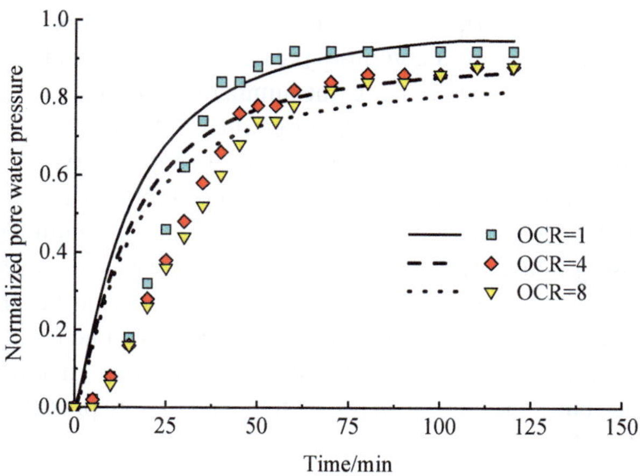

Fig. 4.3 Simulation of pore pressure variation with time during undrained heating

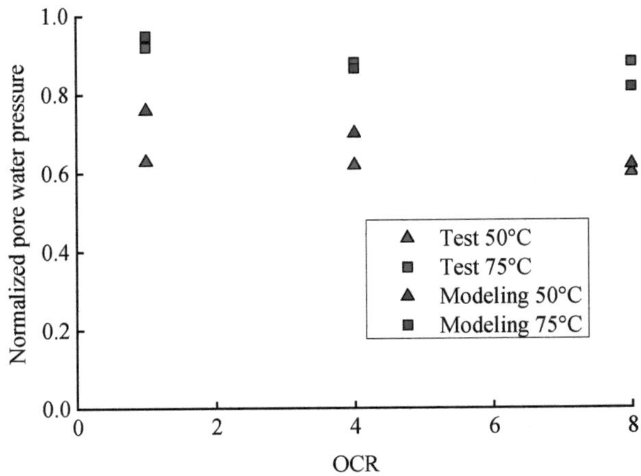

Fig. 4.4 Simulation of pore water pressure change during undrained heating

pressure caused by heating also increases. When heated to a certain temperature and subjected to isothermal solidification, the pore pressure will decrease during the undrained cooling process and gradually drop below zero. The maximum negative pore pressure generated will increase with the decrease of moisture content or the increase of thermal load. When isothermal consolidation is carried out at ambient temperature after a hot cold cycle, a phenomenon of "water absorption" will occur, and the pore water pressure will begin to rise, eventually stabilizing near zero. As the number of hot cold cycles increases, the pore water pressure generated by heating shows a decreasing trend, which is more pronounced under higher confining pressure conditions. And with the increase of the number of hot cold cycles, the soil sample shows a significant decrease in volume, indicating the existence of significant irreversible thermal consolidation deformation, that is, the occurrence of overconsolidation effect.

Figure 4.5 shows the variation of normalized pore water pressure with the number of hot and cold cycles at a confining pressure of 50 kPa and temperature rise/drop amplitudes of 25 °C and 50 °C. As the number of hot and cold cycles increases, the pore pressure generated during heating gradually decreases, and the larger the heating amplitude, the greater the pore pressure generated. When cooling down, negative pore pressure is generated. Similarly, as the number of cycles increases, the absolute value of the negative pore pressure does not change significantly and shows a slight decreasing trend. As the number of cycles increases, the overconsolidation effect also increases, and the pore pressure generated during the heating/cooling process decreases, similar to the experimental results obtained in Fig. 4.3 above. The simulation results have a certain deviation from the experimental results when cooled by 50 °C, but overall they can reflect the trend of pore pressure variation with the number of cycles.

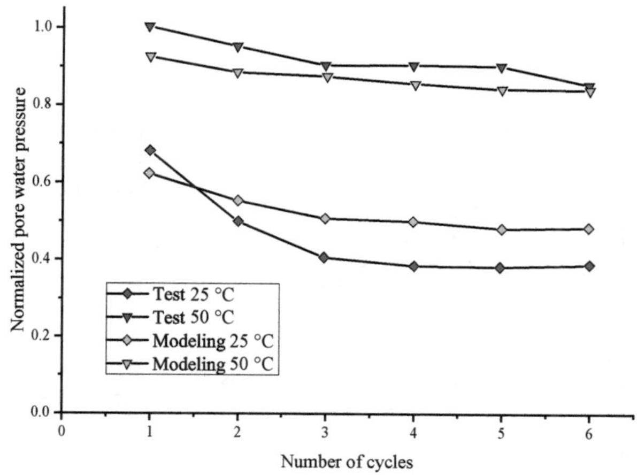

(a) Heating up

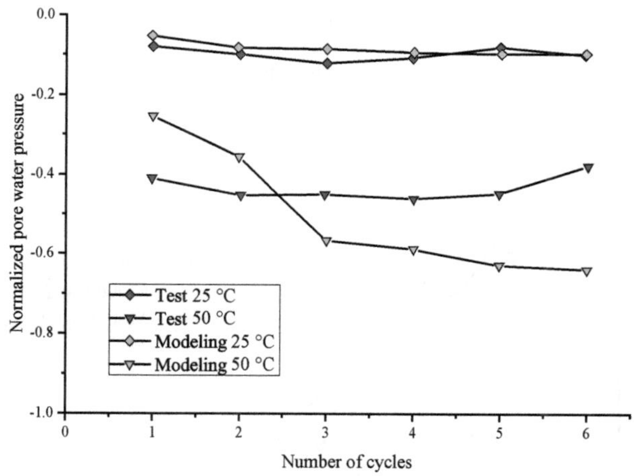

(b) Cool down

Fig. 4.5 Variation of pore water pressure with the cyclic numbers of the thermal loading caused by heating/cooling under confining pressure of 50 kPa

Figure 4.6 shows the variation of thermal strain with the number of thermal load cycles under a confining pressure of 50 kPa during heating/cooling. Due to the increase of overconsolidation effect caused by cycling, the thermal shrinkage of the soil sample decreases, while the thermal expansion relatively increases. Overall, it can be seen that the thermal strain generated by heating/cooling decreases with the increase of the number of cycles. When the temperature was raised to 25 °C, thermal expansion occurred after 4 cycles, while no thermal expansion occurred when the temperature was raised to 50 °C. This may be because when the heating amplitude was large, the soil samples with more cycles had already transitioned from

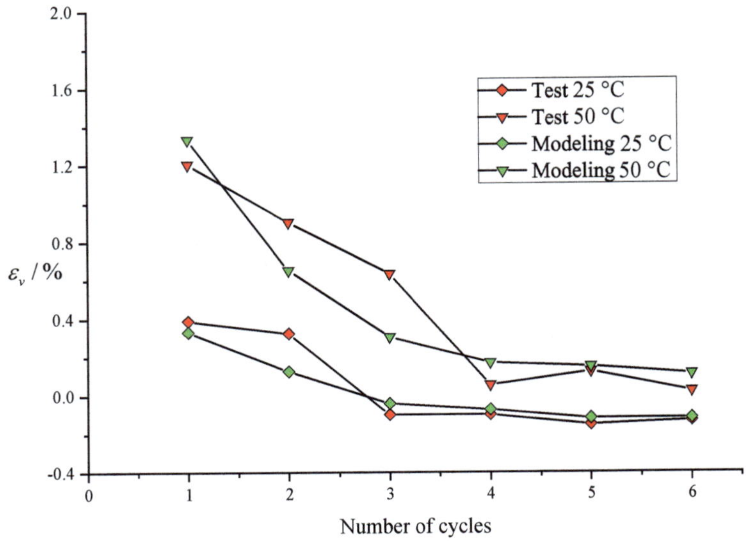

(a) Heating up

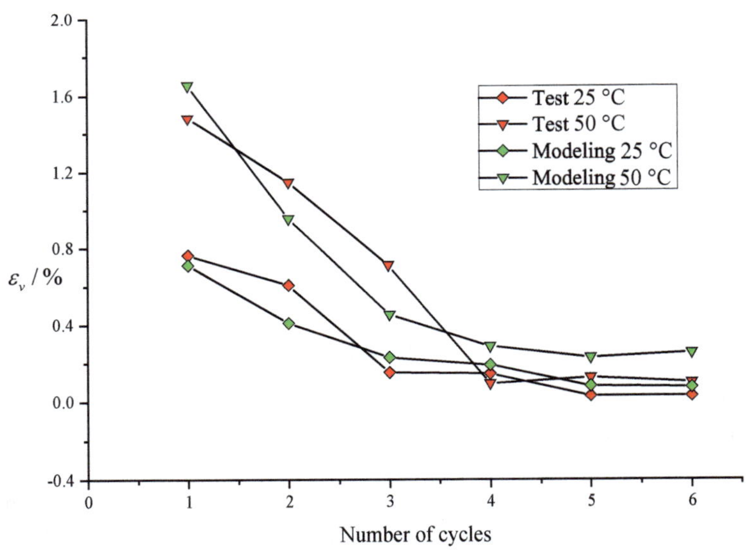

(b) Cool down

Fig. 4.6 Variation of thermal consolidation volume strain with the cyclic numbers of the thermal loading caused by heating/cooling under confining pressure of 50 kPa thermal expansion to thermal shrinkage, as shown in Fig. 4.2. As mentioned earlier, the thermal strain caused by cooling can be regarded as an elastic change, which also decreases with the increase of the number of cycles. It can be seen that the simulation results can reflect the variation law of thermal strain with the number of cycles.

4.4 Summary

In order to verify the theoretical constitutive model proposed in Chap. 2, this chapter conducted thermal solidification tests on saturated silty clay considering overconsolidation ratio under different temperature rise and fall paths using a self-developed sample device. The experimental research showed that:

1. The greater the increase or decrease in temperature without drainage, the greater the induced pore water pressure. The maximum pore water pressure achieved by samples with different overconsolidation ratios does not differ significantly when heated, but during the cooling process, the negative pore pressure generated tends to increase with the increase of overconsolidation ratio.
2. The larger the heating or cooling amplitude, the greater the strain of the consolidated body, and the strain of the consolidated body decreases exponentially with the increase of the overconsolidation ratio. After a heating cooling cycle, the overall strain also decreases with the increase of the overconsolidation ratio.
3. The more temperature levels are set during the heating or cooling process, the greater the corresponding consolidated strain, and as the overconsolidation ratio increases, the difference in strain between different levels decreases.

In addition, based on the basic framework of the unsaturated soil thermal hydraulic coupling model proposed in Chap. 2, for the special case of saturated clay, the model is degraded to the level of saturated soil, and a constitutive model that can describe the thermal hydraulic coupling characteristics of saturated clay is established. The ability of the proposed model to describe the thermal consolidation characteristics of saturated clay was verified through the comparison and analysis of the thermal consolidation test results of saturated clay considering the overconsolidation effect in this chapter, as well as the thermal response test results of saturated clay under cyclic temperature load, with the simulation results of the model.

References

1. Abuel-Naga HM, Bergado DT, Ramana GV, Grino L, Rujivipat P, Thet Y (2006) Experimental evaluation of engineering behavior of soft Bangkok clay under elevated temperature. J Geotech Geoenviron 132:902–910
2. Campanella RG, Mitchell JK (1968) Influence of temperature variations on soil behavior. J Soil Mech Found Div 94:709–734
3. Kuntiwattanakul P, Towhata I, Ohishi K, Seko I (1995) Temperature effects on undrained shear characteristics of clay. Soils Found 35(1):147–162
4. Abuel-Naga HM, Bergado DT, Soralump S, Rujivipat P (2005) Thermal consolidation of soft Bangkok clay. Lowland Technol Int 7(1):13–21
5. Demars KR, Charles RD (1982) Soil volume change induced by temperature cycling. Can Geotech J 19(2):189–194
6. Sultan N, Delage P, Cui YJ (2002) Temperature effects on the volume change behaviour of Boom clay. Eng Geol 64(2–3):135–145
7. Cui YJ, Le TT, Tang AM, Delage P, Li XL (2009) Investigating the time dependent behaviour of Boom clay under thermo-mechanical loading. Géotechnique 59(4):319–329
8. Hueckel T, Pellegrini R (1992) Effective stress and water pressure in saturated clays during heating-cooling cycles. Can Geotech J 29(6):1095–1102
9. Romero E, Villar MV, Lloret A (2005) Thermo-hydro-mechanical behaviour of two heavily overconsolidated clays. Eng Geol 81(3):255–268
10. Cui YJ, Tang AM (2013) On the chemo-thermo-hydro-mechanical behaviour of geological and engineered barriers. J Rock Mech Geotech Eng 5(3):169–178
11. Monfared M, Delage P, Sulem J, Mohajerani M, Tang AM, De Laure E (2011) A new hollow cylinder triaxial cell to study the behavior of geo-materials with low permeability. Int J Rock Mech Min Sci 48(4):637–649
12. Zhang ZC (2017) A thermodynamics-based theory for the thermo-poro-mechanical modeling of saturated clay. Int J Plast 90:164–185
13. Bai B, Shi XY (2017) Experimental study on the consolidation of saturated silty clay subjected to cyclic thermal loading. Geomech Eng 12(4):707–721
14. Bai B, Su ZQ (2012) Thermal responses of saturated silty clay during repeated heating-cooling processes. Transp Porous Media 93(1):1–11
15. Bai B (2013) Thermal response of saturated porous spherical body containing a cavity under several boundary conditions. J Therm Stresses 36(11):1217–1232
16. Abuel-Naga HM, Bergado DT, Lim BF (2007) Effect of temperature on shear strength and yielding behavior of soft Bangkok clay. Soils Found 47(3):423–436
17. Laloui L, Cekerevac C (2008) Numerical simulation of the non-isothermal mechanical behaviour of soils. Comput Geotech 35:729–745
18. Abuel-Naga H, Bergado D, Bouazza A, Ramana G (2007) Volume change behaviour of saturated clays under drained heating conditions: experimental results and constitutive modeling. Can Geotech J 44(8):942–956

Chapter 5
Model Description of the Mechanical Properties and Temperature Effects of Fine-Grained Gassy Soils

5.1 Introduction

In some engineering projects near the ocean, rivers, lakes, etc., a special type of gas often develops in soft soil strata, which is called "shallow gas" in engineering geology. These gases are mainly methane, with small amounts of nitrogen and carbon dioxide. They may come from biological decomposition or thermogenic processes [1], and are mainly formed in fine-grained soils (clay, silt soil) under the seabed [2, 3]. In the era of promoting the use of new energy, shallow gas can be regarded as an important resource. However, in engineering construction, it is a potential geological hazard. In engineering projects in soft soil areas, geotechnical engineering problems caused by shallow gas often occur, and in severe cases, major disasters may even be triggered. In seabed sediments, gas exists in three forms: dissolved in the liquid phase, undissolved gas in the form of voids, and in the form of natural gas hydrates [5]. The release of these undissolved gases, the exsolution of dissolved gases, and the gas release caused by the decomposition of natural gas hydrates will have a significant impact on the engineering properties of the marine sediments that carry them. Liu et al. [6] conducted an experimental study on the bubble growth characteristics of natural gas-bearing sediments (sand and clay) through high-resolution X-ray scanning technology, and the results are shown in Fig. 5.1. For fine-grained soils, with the increase of time, the accumulation of bubbles will generate relatively large gas cavities, while for sandy soils, the gas mainly exists in the form of small bubbles [7]. At present, the engineering geological research on gas-bearing soils mainly focuses on sand layer foundations. Many scholars [7–10, 35] have studied the experimental characteristics of gas-bearing sand (with small bubbles in pore water). There are relatively few reports on experimental research data of gas-bearing fine-grained soils (soils with large bubbles) [2, 11–13], possibly due to the long experimental period caused by the low permeability of clay and the difficulty of sample preparation. Gas-bearing fine-grained soil strata mainly include

G. Yang, *Multi-field Coupled Constitutive Model for Soils Based on Granular Thermodynamics Theory*, https://doi.org/10.1007/978-981-95-0249-3_5

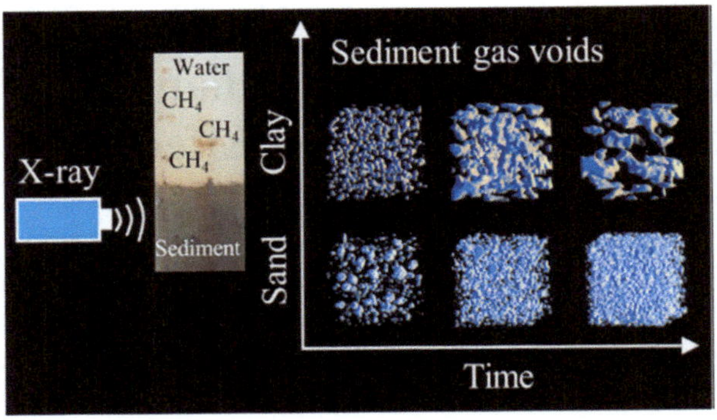

Fig. 5.1 Time-dependent changes of air bubbles in gassy sand and clay [6]

silt, silty clay, and silty (silty fine-grained soil). These strata are rich in a large amount of organic matter. Under suitable conditions, these organic matters are decomposed and fermented by biological action to form shallow gas. After the formation of shallow gas, apart from the part that escapes to the outside, the remaining gas is generally adsorbed by the soil layer and dissolved in the pore water of the soil. Only when the adsorption capacity of the soil layer and the dissolution effect of the pore water reach saturation will free gas be formed and exist in the soil layer in the form of bubbles.

At present, the research on gas-bearing fine-grained soils mainly remains at the experimental stage. Due to the special occurrence conditions of soil, it is difficult to obtain undisturbed intact soil. The gas-bearing soil samples used in experiments are generally remolded soil samples, which are prepared by adding zeolite or methane-producing microorganisms to the soil samples [14]. Due to the difficulties in sample preparation and the operation of experiments, there are relatively few existing experimental studies. These mainly include research on the consolidation and shear characteristics of gas-bearing fine-grained soils at normal temperature [2, 11–19]. The experimental results of the mechanical properties of gas-bearing fine-grained soils can be summarized as follows:

(1) Under monotonic loading, the change of the gas part is related to the change of total stress, and the change of the saturated soil matrix part is related to the consolidation stress (total stress minus pore water pressure).
(2) The existence of large bubbles may cause an increase or decrease in the undrained shear strength, depending on the applied confining pressure and the initial effective stress of the saturated soil matrix part. Moreover, the initial pore water pressure has a significant impact on the undrained shear strength.
(3) The pore gas pressure is always greater than the pore water pressure. The surface tension at the water-gas interface limits the difference between the gas pressure and the water pressure. At the same time, the expansion or contraction of the gas cavity limits the difference between the gas pressure and the total stress.

(4) The failure of soil caused by the precipitation and expansion of gas will lead to an increase in the isotropic compressibility of the soil, and a decrease in the pre-consolidation pressure and the peak undrained strength.

Regarding theoretical research, Wheeler [16], based on the proposed large-bubble model, gave the upper and lower bounds of the undrained shear strength of gas-bearing fine-grained soils and the expression of the elastic modulus. Pietruszczak and Pande [21], and Grozic et al. [22], respectively, proposed constitutive models that can reflect the undrained compression characteristics of gas-bearing fine-grained soils. Both models are based on the concept of critical state and assume that the change of the gas phase only affects the volume change of gas-bearing fine-grained soils. Sultan and Garziglia [23], through the expansion of the Cambridge model, introduced a gas failure parameter and proposed a constitutive model that can reflect the influence of the growth and aggregation of bubbles on gas-bearing fine-grained soils.

In summary, there is very little research on the theoretical models of gas-bearing fine-grained soils in the existing literature. The few existing models only focus on a single characteristic of gas-bearing fine-grained soils. Moreover, neither the above-mentioned experimental nor theoretical studies have addressed the temperature characteristics of gas-bearing soils. Global warming, natural gas exploitation, and the construction of some underground projects will all affect the temperature of the occurrence conditions of gas-bearing soils, thus influencing their engineering mechanical properties. Therefore, the study of the temperature characteristics of gas-bearing soils is essential. Based on the concept of the large-bubble model proposed by Wheeler, that is, gas-bearing fine-grained soils are composed of closed large bubbles and the surrounding saturated soil matrix, this chapter combines the thermo-mechanical coupling model of saturated soil given in Chap. 4 and applies it to the saturated soil matrix part of gas-bearing fine-grained soils. Considering the influence of the gas phase on the deformation of the soil skeleton, a thermo-mechanical model that can describe the mechanical properties and temperature effects of gas-bearing fine-grained soils is developed. This model can well describe the consolidation and shear characteristics of gas-bearing fine-grained soils, and the ability to describe its temperature characteristics is also explored.

5.2 Thermodynamic Constitutive Model of Fine-Grained Gassy Soils

Thomas [17] found that the change in the gas phase volume $e_g + He_l$ is related to the change in the total stress σ, and it has a linear relationship with $\ln\sigma$. This indicates that the change in the gas phase is mainly controlled by the total stress [11, 17]. Meanwhile, the void ratio e_l of the saturated soil matrix is related to the change in the effective-stress-like σ'' (the difference between the total stress and the pore water pressure), and it has a linear relationship in the $e_l - \ln\sigma''$ relationship. It is independent of the gas content and is similar to the change in the normal compression curve of saturated soil. Therefore, it can be considered that the mechanical properties of the saturated soil matrix part are mainly controlled by σ'' [2, 11, 17, 19].

The gas-bearing fine-grained soil is composed of closed air bubbles and the surrounding saturated soil matrix. Then the gas void ratio e_g, the void ratio e_l of pore water in saturated soil, and the total void ratio e can be expressed respectively as:

$$e_g = V_g / V_s = (1 - s_r)e \tag{5.1a}$$

$$e_l = V_l / V_s = s_r e \tag{5.1b}$$

$$e = e_g + e_l \tag{5.1c}$$

where V_g, V_l, and V_s are the volumes of the gas phase, liquid phase, and solid phase, respectively; e is the total void ratio of the gas-bearing soil; and s_r is the degree of saturation of the gas-bearing soil.

5.2.1 *Characteristics of the Gas Phase*

The gas within the gas-containing soil can dissolve in the pore fluid or be released from the pore fluid, and it follows both Boyle's Law and Henry's Law. Combining the two laws, we obtain:

$$(e_g + he_l)(u_g + p_a) = \text{constant} \tag{5.2}$$

where p_a is the atmospheric pressure, approximately 101.325 kPa; u_g is the pore gas pressure; h is the Henry's solubility coefficient; and the solubility coefficient of methane in water at 20 °C is 0.0333 [24].

Assuming the gas follows the ideal gas law, the equation is:

$$u_g V_g = n_{mol} RT \tag{5.3}$$

where n_{mol} is the number of moles of the gas; R is the constant; T is the absolute temperature.

When the temperature and gas content are constant, Eq. (5.3) degenerates into Eq. (5.2). The ideal gas law indicates that, although the gas content is the same, the value of $u_g V_g$will vary at different temperatures. This should be taken into account when studying the effect of temperature.

Normalization of the volume change in the gas phase yields:

$$\frac{e_g + he_l}{e_{g0} + he_{l0}} = \frac{p_a}{p_a + u_g} \tag{5.4}$$

Since gas in gas-bearing fine-grained soils exists in the form of large, discrete gas bubbles, it is difficult to directly measure the gas pressure. Thomas [25] conducted linear loading consolidation tests under drainage conditions to measure the changes in total stress, total volume, and pore water pressure of gas-bearing soil samples. Based on Boyle's law and Henry's law, he described the variation in gas pressure, as shown in Fig. 5.2.

In the first stage of linear loading (AB), as the total stress increases linearly, the gas pressure also increases approximately linearly. The gap between the total stress and gas pressure gradually widens, while the pore water pressure initially increases and then decreases.

In the second stage (BC), when the total stress reaches point B and remains constant, the gas pressure also remains constant, and the pore water pressure gradually decreases.

In the third stage (CD), during the initial decrease of total stress (CE), the gas pressure remains unchanged. After point E, the total stress continues to decrease, and the gas pressure also begins to decrease. The curves of their decrease are similar, and during this stage, the pore water pressure remains constant.

Overall, during the entire loading-unloading process, the relationship between gas pressure and total stress is approximately linear, despite some special phenomena observed during unloading (when total stress decreases but gas pressure remains unchanged in stage CE).

Based on the experimental observations and the analysis of the soil continuum model, Wheeler and Sham [5] proposed two theoretical hypotheses for independent gas pressure limits: the surface tension effect limit and the cavity expansion-contraction limit. These two gas pressure limit hypotheses suggest that when the gas pressure falls within the range of these two limits, it will not be affected by the pore

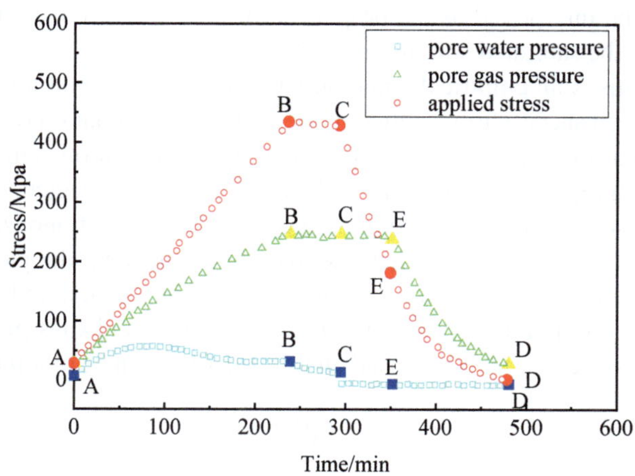

Fig. 5.2 The variation of vertical stress, pore water pressure, and pore gas pressure with time in linear loading consolidation tests [12]

water pressure (in the AE stage of Fig. 2.1) or by the total stress (in the CE stage of Fig. 2.1).

For the first limit hypothesis, the surface tension effect limits the difference between the gas pressure u_g and the pore water pressure u_l. At any given moment, ug is greater than u_l, corresponding to the lower limit of the surface tension effect, $u_g > u_l$. Likewise, there is no evidence to suggest that ug has reached the upper limit of surface tension, possibly because $2T_s/R_c$ is very large, where Ts is the surface tension and R_c is the radius of curvature. It can be observed that the limitation of the surface tension effect is almost always satisfied. Therefore, it can be assumed that at low pore pressures, ug is almost unaffected by changes in pore water pressure (the AE stage in Fig. 5.2), because these changes are absorbed into the changes in the curvature of the curved liquid surface [12].

For the second limit hypothesis, based on the assumption that the saturated soil matrix surrounding the cavity behaves as an elastic-plastic continuum, the expansion or contraction of the pore gas is calculated based on the difference between the gas pressure and the average total stress applied to the soil. The CE stage in Fig. 5.2 provides strong evidence supporting this hypothesis, where the gas pressure is unaffected by changes in total stress within this limit. This may be due to the fact that the stiffness of the saturated soil matrix is much greater than that of the gas [12].

At the same time, Wheeler and Sham [5] also pointed out that while the experimental results support some aspects of the theoretical relationships, certain facts are not entirely consistent with the theoretical analysis, and full validation requires further experimental research.

Based on the experimental observations, Thomas [17] suggested that during consolidation tests, the relationship between gas pressure and total stress can be approximated as a linear relationship, and provided an empirical formula:

$$u_g = a_g \sigma \tag{5.5}$$

where a_g can be interpreted as the proportion of total stress transmitted to the gas phase through the saturated soil matrix $a_g \leq 1$.

Although this is an empirical equation based on experimental results, this relationship has been successfully applied in the numerical simulation of consolidation tests on gas-bearing fine-grained soils [25]. During consolidation compression tests, pore pressure dissipates continuously, and its effect can be neglected. However, during undrained shear tests, pore water pressure accumulates continuously, which passively drives the gas pressure to increase, thereby influencing the gas pressure. At this point, the effect of pore water pressure can no longer be ignored. Furthermore, in gas-bearing fine-grained soils, the pore gas pressure is always greater than the pore water pressure [12], $u_g > u_l$. Based on this, Eq. (2.7) can be modified as follows:

$$u_g = a_g \sigma + u_l \tag{5.6}$$

For consolidation tests, since pore pressure continuously dissipates ($u_l \rightarrow 0$), Eq. (5.6) degenerates into the form of Eq. (5.5). At the same time, Eq. (5.6) also ensures that the pore gas pressure is greater than the pore water pressure. Based on this, in the model proposed in this chapter, the gas pressure is expressed using the empirical Eq. (5.6), and the influence of the value of a_g will also be analyzed later in the text.

5.2.2 *Effective Stress of Fine-Grained Gassy Soil*

In previous research works [2, 11, 12, 19], the mechanical behavior of the saturated soil matrix in gas-bearing fine-grained soils was analyzed based on "effective stress" (the difference between total stress and pore water pressure). To distinguish it from the effective stress in saturated soils, this is referred to as "quasi-effective stress" or "operative stress." In fact, simply using the difference between total stress and pore water pressure to represent the effective stress of the saturated soil matrix is not sufficiently accurate. Wheeler [11] pointed out that the behavior of the saturated soil matrix in gas-bearing fine-grained soils should depend on the difference between the total stress and pore water pressure within the matrix. The saturated soil matrix only carries a large portion of the total stress, but not the entire stress. On this basis, a mechanical equilibrium analysis of a section of fine-grained gas-bearing soil allows for the determination of the effective stress of the saturated soil matrix.

Consider a section of gas-bearing fine-grained soil, as shown in Fig. 2.2. Let A represent the cross-sectional area, σ represent the total stress acting on the section, A_m represent the area occupied by the saturated soil, σ_m represent the stress borne by the saturated part, A_{gi} represent the area occupied by a single enclosed gas bubble, and $\sum A_{gi}$ represent the total area of the gas. Thus, the mechanical equilibrium on the section can be expressed as:

$$\sigma A = \sum u_{gi} A_{gi} + \sigma_m \left(A - \sum A_{gi} \right) - T_s \sum l_i \sin \theta_i \tag{5.7}$$

where T_s is the surface tension at the water-gas interface, l_i is the perimeter of the intercepted gas bubble, and θ_i is the contact angle between the solid and liquid phases.

In fact, the interface between pore water and pore gas is composed of many small menisci, as shown in Fig. 5.3. It can be assumed that the radius of the bubble is independent of the surface tension between water and gas, and thus independent of the pressure of the bubble [17]. Since both the radius of curvature of the water-gas meniscus and the surface tension are extremely small, the influence of surface tension on the mechanical properties of gas-bearing soil can be neglected [11, 17].

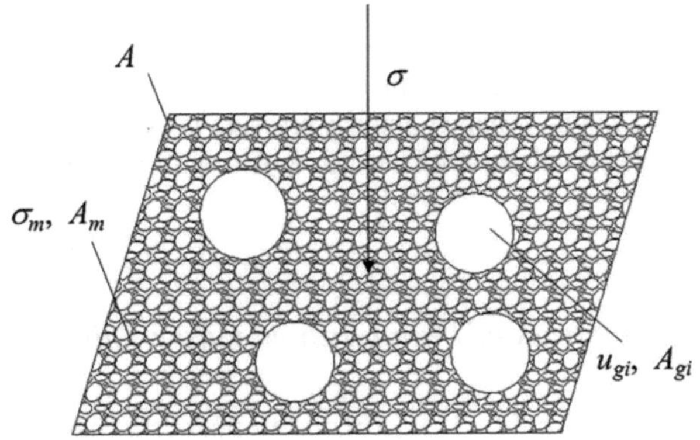

Fig. 5.3 Composition and stress of fine-grained gassy soil section

Therefore, the last term in Eq. (5.7) can be neglected, and we have:

$$\sigma_m = \sigma \frac{A}{A - A_g} - u_g \frac{A_g}{A - A_g} \tag{5.8}$$

where $A_g = \sum A_{gi}$ represents the total area of the section occupied by the gas.

The gas volume fraction can be expressed as:

$$n_g = \frac{A_g}{A} = \frac{e_g}{1 + e} \tag{5.9}$$

By combining Eq. (5.6), we can obtain:

$$\sigma_m = \frac{(1 - a_g n_g)}{1 - n_g} \sigma - \frac{n_g}{1 - n_g} u_l \tag{5.10}$$

Therefore, let's redefine the effective stress.

$$\sigma'' = \sigma_m - u_l = \frac{(1 - a_g n_g)}{1 - n_g} \sigma - \frac{1}{1 - n_g} u_l \tag{5.11}$$

When $n_g = 0$ is the case, $\sigma'' = \sigma - u_l$ is the same as the "effective stress" used in previous studies and degenerates into the form of Terzaghi's effective stress. Thomas [17], following the method of Timoshenko and Goodier [20], idealized gas-bearing fine-grained soil as a hollow elastic sphere and obtained a derivation similar to Eq. (2.10). In addition, Thomas analyzed the effect of the value of a_g on the consolidation curve. The study showed that even with a wide range of values for a_g, the differences in the consolidation curve were minimal. Therefore, an approximate analysis can also be performed using the form of $\sigma'' = \sigma - u_l$. In this chapter, a more precise form, Eq. (5.11), is used (Fig. 5.3).

5.2.3 Influence of Gas-Phase Deformation on Saturated Soil Matrix

Suppose the gas-bearing fine-grained soil is regarded as a system. The deformation rate of the gas, $d_t\varepsilon_v^g$ (with no shear strain), should also be a dissipative force, and its corresponding dissipative flow is the viscous stress of the gas, σ_{ij}^{vG}. Therefore, in the original migration coefficient model of saturated soil (Eq. 4.10), the dissipative term of the gas phase should be added. Since the compressibility of the gas is extremely high, the gas-bearing soil will deform even under undrained loading conditions. So, the change in gas volume will affect the volumetric strain of the soil skeleton to a certain extent [21, 22]. For example, in the initial stage of undrained shearing, the gas volume is easily affected by disturbance and changes, thus causing the overall volume to change. At this time, the undrained effective stress path of the gas-bearing soil is often closer to the drained path [2]. Therefore, in the dissipative relationship, to consider the influence of the gas phase on the deformation of the soil skeleton, a correlation coefficient τ_{ijkl} is introduced, and then the migration coefficient model can be expressed as:

$$\begin{bmatrix} d\varepsilon_{ij}^p \\ I_g \\ q_i \\ \sigma_{ij}^{vS} \\ \sigma_{ij}^{vL} \\ \sigma_{ij}^{vG} \end{bmatrix} = \begin{bmatrix} \eta_{ijkl} & 0 & 0 & 0 & \tau_{ijkl} \\ 0 & \gamma & 0 & 0 & 0 \\ 0 & 0 & \kappa_{ik} & 0 & 0 \\ 0 & 0 & 0 & \lambda_{ijkl}^S & 0 \\ 0 & 0 & 0 & 0 & \lambda_{ijkl}^L \\ \tau_{ijkl} & 0 & 0 & 0 & \lambda_{ijkl}^G \end{bmatrix} \begin{bmatrix} \pi_{kl} \\ T_g \\ \nabla_k T \\ d\varepsilon_{kl} \\ \psi_{kl}^L \\ \psi_{kl}^G \end{bmatrix} \tag{5.12}$$

Then, the expression of the plastic strain rate considering the influence of the gas phase can be obtained:

$$d_t\varepsilon_{ij}^p = \eta_{ijkl}\pi_{kl} + \tau_{ijkl}\psi_{kl}^g \tag{5.13}$$

Since the shear strain of the gas is not considered, and the change in gas volume only affects the volumetric strain of the soil skeleton, the above-mentioned formula can be simplified as:

$$d_t\varepsilon_{ij}^p = \eta_v\left(T_g\right)^\alpha\left(\varepsilon_v^e + \beta_T\Delta T\right)\delta_{ij} + \eta_s\left(T_g\right)^\alpha e_{ij}^e + \tau_v\delta_{ij}d_t\varepsilon_v^g \tag{5.14}$$

where, τ_v is the coupling coefficient, which reflects the influence of gas on the deformation of the soil skeleton.

For the saturated soil matrix part, it can be regarded as a completely saturated soil body alone. Therefore, it can be controlled by the thermo-mechanical model of saturated soil given in Chap. 4. On this basis, considering the physical state equation of gas (Eq. (5.3)), the form of effective stress of gas-bearing fine-grained soils (Eq.

(5.11)), and the influence of gas volume change on the volumetric strain of the soil skeleton (Eq. (5.14)), a thermo-mechanical model for gas-bearing fine-grained soils can be given. This model can describe the basic mechanical properties and thermo-mechanical coupling properties of gas-bearing fine-grained soils.

5.3 Simulation Analysis of the Mechanical Properties of Fine-Grained Gassy Soils

5.3.1 Simulation of the Experiments by Wheeler [15] and Thomas [17]

Wheeler [15] used chemical techniques to generate large methane bubbles inside remolded silty loam, prepared gas-bearing soil samples, and carried out a series of triaxial undrained shear tests on the soil samples to study the undrained shear strength characteristics and the stress-strain characteristics before failure of gas-bearing soils. On the basis of Wheeler's research, Thomas [17] conducted an isotropic consolidation compression test and a triaxial isotropic compression test on the same type of gas-bearing soil to study the influence of gas on the consolidation and compression characteristics of gas-bearing soils. The experimental data of the two are used to verify the ability of the model given in this chapter to describe the consolidation and shear characteristics of gas-bearing fine-grained soils. Since the experimental data were all obtained at normal temperature, the influence of temperature is not considered. All the involved model parameters can be determined according to the method given in Sect. 3.1 of Chap. 3. τ_v can be determined through the undrained shear test of gas-bearing fine-grained soils and is related to the effective stress path, as shown in Fig. 5.7 in the following text. The specific model parameter values are shown in Table 5.1.

(1) Consolidation and Compression Characteristics

For the isotropic consolidation and compression test $\left(\varepsilon_s^e = 0,\ u_l = 0\right)$, without considering the influence of temperature, by differentiating Formulas (4.22) and (5.11), we can obtain:

Table 5.1 Model parameters

Basic parameters			
G_S	2.64	λ	0.59
a_g	0.7	M	1.49
Thermodynamic parameters			
c_1	0.06	$(c_2)_{sat}$	610.23 s^2
c_3	7300 s^2	c_5	76.30 kg/(m^3·s)
B_0	2.10 × 10^5 kPa	c	0.028
τ_v	0.004		

$$d_t p' = \frac{\rho_L G_s p'}{\lambda(s)\rho_d} d_t \varepsilon_v + 1.5B\left(\varepsilon_v^e + \beta_T \Delta T\right)^{0.5} d_t \varepsilon_v^e \tag{5.15}$$

$$d_t p' = \frac{1 - a_g n_g}{1 - n_g} d_t \sigma + \sigma \frac{1 - a_g}{\left(1 - n_g\right)^2} d_t n_g \tag{5.16}$$

The proposed gas-bearing soil model is used to simulate and calculate the isotropic consolidation and compression tests A1/A5/A6 carried out by Thomas [17]. The initial saturations s_{r0} are 1, 0.905, and 0.817, respectively, and the test stress change rate $d_t\sigma$ is 1.707 kPa/min. The comparison between the relationships of the pore-water ratio, pore-gas ratio, and total void ratio with the effective stress obtained from the model calculation and the test results is shown in Fig. 5.4. It can be seen that the test data and the simulation results are in good agreement. For soil samples with different gas contents, the variation laws of the pore-water ratio e_l are almost the same and tend to the consolidation curve of saturated soil samples, as shown in Fig. 5.4a. This indicates that in gas-bearing fine-grained soils, the change of pore water is mainly controlled by the effective-like stress and has nothing to do with the gas content. For soil samples with different gas contents, the normalized change of the pore-gas ratio e_g/e_{g0} is slightly different. When calculating with the model, $a_g = 0.7$ is set, and the calculation results are just at an average value position, as shown in Fig. 5.4b. The change of the total void ratio e is more obviously affected by the change of the gas content. As shown in Fig. 5.4c, the larger the gas content, the steeper the slope of the compression curve. It can be seen that the proposed model can well simulate the consolidation and compression characteristics of gas-bearing fine-grained soils.

In the model, it is generally believed that changes in the gas phase are approximately linearly related to changes in total stress. The value of the correlation coefficient a_g has a significant impact on simulating changes in the pore gas ratio e_g, as shown in Fig. 5.5. It can be seen that with the increase in gas content, the value of a_g also increases. That is to say, the higher the gas content, the closer the gas pressure value is to the external total stress value, which is consistent with common sense.

(2) Undrained Shear Characteristics

The occurrence environment of gas-bearing fine-grained soils is generally located under lakes or oceans, approximately in an undrained condition. Therefore, the research on the shear characteristics of gas-bearing fine-grained soils mainly focuses on their undrained shear characteristics. Wheeler [15] carried out a series of undrained shear tests on gas-bearing fine-grained soil samples with different initial conditions. The research found that the shear strength of gas-bearing soil samples has uncertainty with the increase of gas content, either increasing or decreasing, which is related to the initial conditions. For example, when $\mu_l = 0$ and $\sigma_3 = 400$ kpa, the increase of gas content will increase the shear strength; while when $\mu_l = 100$ kpa and $\sigma_3 = 200$ kpa, the increase of gas content will decrease the shear strength. This

Fig. 5.4 Numerical simulations of consolidation tests: (**a**) water void ratio; (**b**) normalized gas void; (**c**) total void ratio

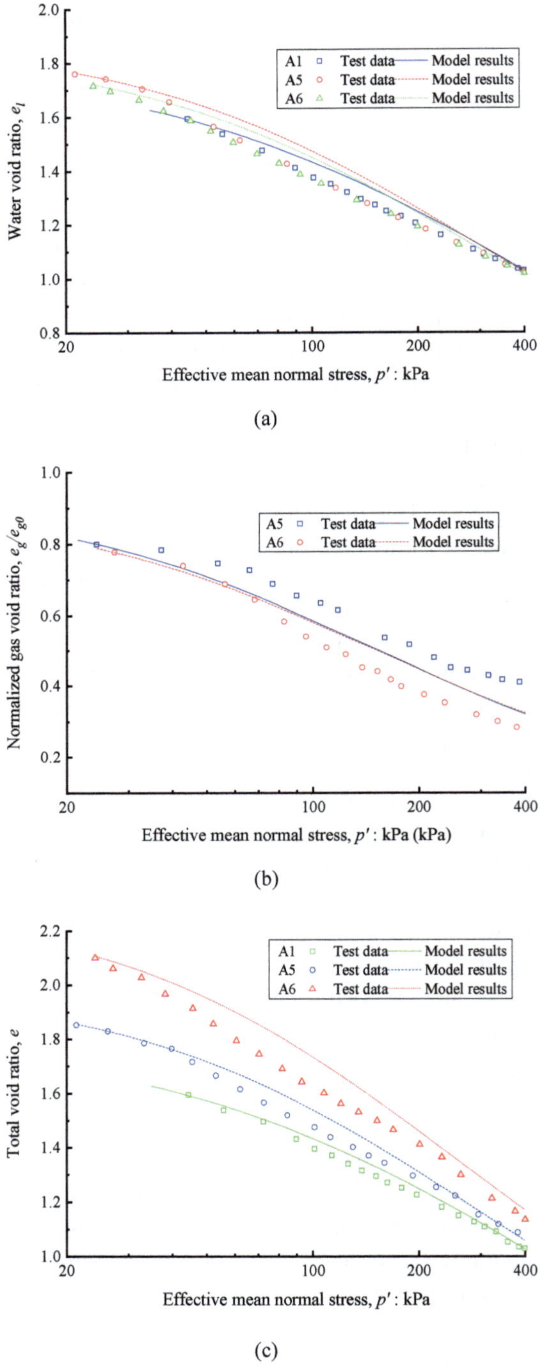

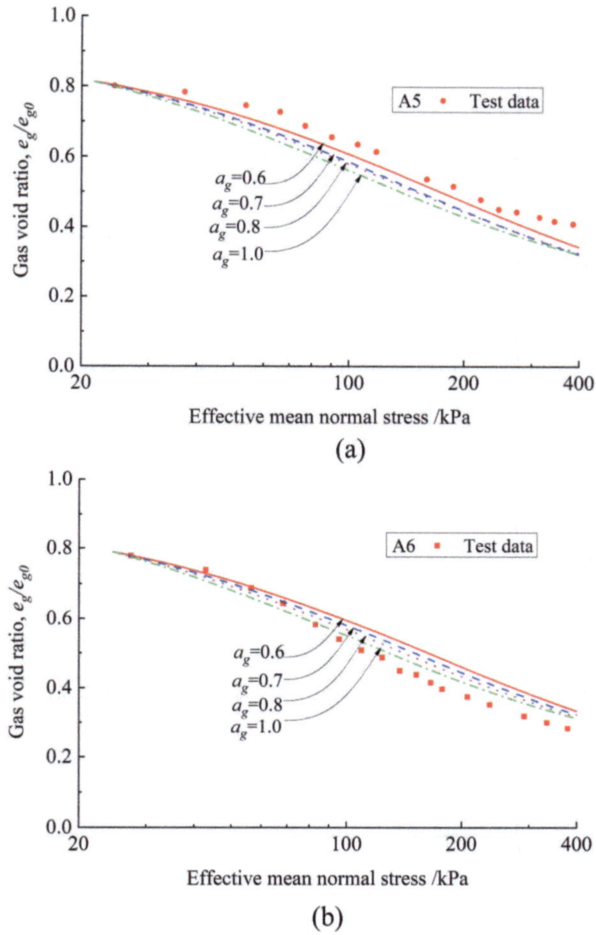

Fig. 5.5 Effect of a_g on normalized gas void ratio e_g/e_{g0}: (**a**) Test A5, s_{r0} = 0.905; (**b**) Test A6, s_{r0} = 0.817

shows that the shear strength is jointly controlled by the effective-like stress and the initial confining pressure [2, 12, 13]. However, the determination of whether the shear strength increases or decreases can usually only be determined through experiments, and there is no specific dividing point. The general rule found in the experiments is that when the initial pore-water pressure is small, the shear strength increases with the increase of gas content, and vice versa.

In this model, the shear strength q is mainly controlled by the critical elastic shear strain $(\varepsilon^e_s)_{cr}$ and c (Eq. 4–23). When the critical state is reached, $\left(\varepsilon_s^a\right)_{cr} = C_2^{0.5}$ can be obtained. Also, since $q/p' = M$ under the critical state, $c_2^{0.5}c$ = constant can be derived, that is, the magnitudes of both $(\varepsilon^e_s)_{cr}$ and c are related to c_2. Therefore, the value of c_2 is related to the shear strength. Under the saturated state, the value of $(c_2)_{sat}$ can be calculated. As the gas content increases, the value of c_2 also changes,

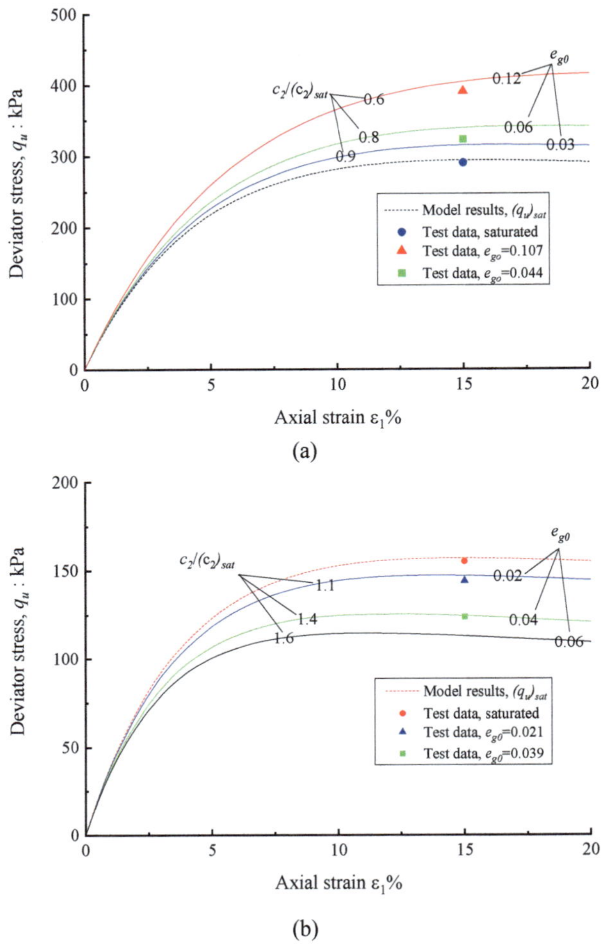

Fig. 5.6 Stress–strain relationship of fine-grained gassy specimens at different initial conditions: (**a**) the undrained shear strength increases with the increase of gas content at $\mu_l = 0$, $\sigma_3 = 400$ kPa; (**b**) the undrained shear strength decreases with the increase of gas content at $\mu_l = 100$ kPa, $\sigma_3 = 200$ kPa

which can be represented by $c_2/(c_2)_{sat}$. When the increase of gas content increases the shear strength, $c_2/(c_2)_{sat}$ decreases, as shown in Fig. 5.6a. When the increase of gas content decreases the shear strength, $c_2/(c_2)_{sat}$ increases, as shown in Fig. 5.6b. The dotted line represents the saturated soil sample.

The simulation results of the corresponding effective-stress paths for the above two types of specimens with different initial conditions are shown in Fig. 5.7, where the dotted line represents the effective-stress path of the saturated specimen. When the initial conditions are $\mu_l = 0$, $\sigma_3 = 400$ kPa, with the increase of the gas volume fraction $f\left(f = \frac{(1-s_r)e}{1+e}\right)$, the effective-stress path moves upward; when the initial

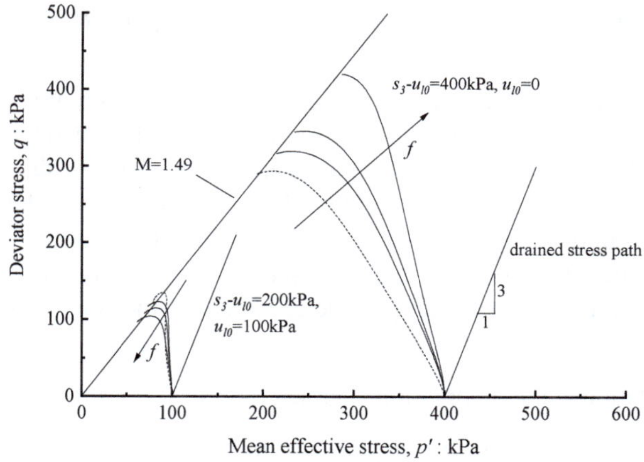

Fig. 5.7 Simulation of effective stress path

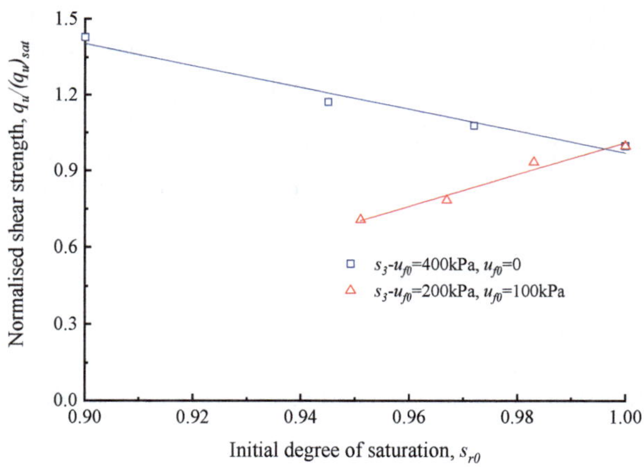

Fig. 5.8 Relation between undrained shear strength and initial saturation of fine-grained gassy soil

conditions are u_l = 100 kPa and σ_3 = 200 kPa, with the increase of the gas volume fraction *f*, the effective-stress path moves downward. It can be seen that this model can reflect the variation characteristics of the undrained shear strength of gas-bearing fine-grained soils.

The relationship between the undrained shear strength of gas-bearing fine-grained soils and the initial saturation is shown in Fig. 5.8. Whether the presence of gas promotes or weakens the shear strength, it almost changes linearly with the undrained shear strength. Based on this, gas-bearing fine-grained soil samples under different initial conditions can be predicted, and it can also provide some reference for certain engineering problems in gas-bearing fine-grained soil strata.

5.3.2 *Simulation of the Experiment by Hong et al. [2]*

Hong et al. [2] carried out a series of undrained shear tests on remolded gas-bearing fine-grained soils under the same effective stress (σ_3'), and studied the consolidation and undrained shear characteristics of gas-bearing fine-grained soils under different initial pore-water pressures u_{l0} (0–1000 kPa). In this chapter, the experimental results under four different initial conditions are selected for simulation and calculation to verify the effectiveness of the proposed gas-bearing fine-grained soil model. The initial conditions and model parameter values of the four tests are shown in Table 5.2.

(1) Consolidation and Compression Characteristics

Before the undrained shear tests are carried out, the four soil samples are first isotropically compressed to 200 kPa under an effective stress of $\sigma_3' = 60$ kPa. The consolidation curves of different initial pore-water pressures are shown in Fig. 5.9. It can be seen that the changes in the pore-water ratio are almost the same, which is consistent with the results obtained above. The change in pore water can be regarded as being controlled only by the effective-like stress, and has nothing to do with the gas content and the magnitude of the initial pore-water pressure. For soil samples with different initial pore-water pressures, the changes in the total void ratio are also different. Similar to Fig. 5.4c, this is mainly caused by the different gas contents of the soil samples. It can be seen that the simulation results are in agreement with the test results.

(2) Undrained Shear Characteristics

Regarding the undrained shear strength of gas-bearing soil samples with different initial pore-water pressures, the situation in Fig. 5.6 above also occurs. When $\mu_{l0} = 0$, the shear strength of the gas-bearing soil is greater than that of the saturated

Table 5.2 Initial conditions and model parameters

σ_3-u_{l0} (kPa)	u_{l0} (kPa)	S_{r0}	e_{g0}
Initial conditions			
200	120	1	0
200	0	0.916	0.131
200	120	0.952	0.071
200	600	0.982	0.026
Basic parameters			
G_S	2.73	λ	0.26
a_g	1.0	M	1.0
Thermodynamic parameters			
c_1	0.10	$(c_2)_{sat}$	734.15 s^2
c_3	6900 s^2	c_5	76.30 kg/(m^3·s)
B_0	9.43×10^6 kPa	c	0.016
τ_v	0.002		

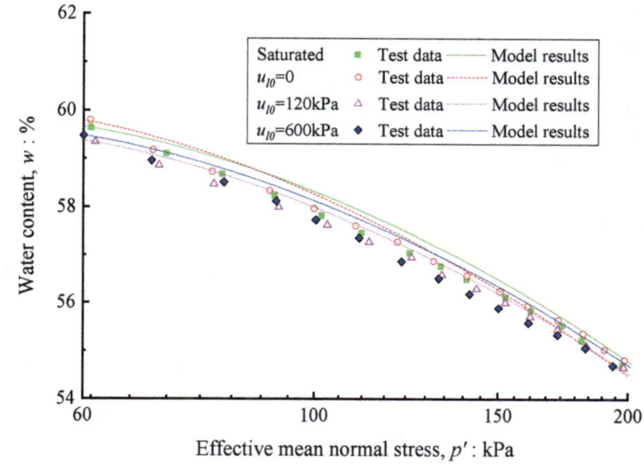

(a)

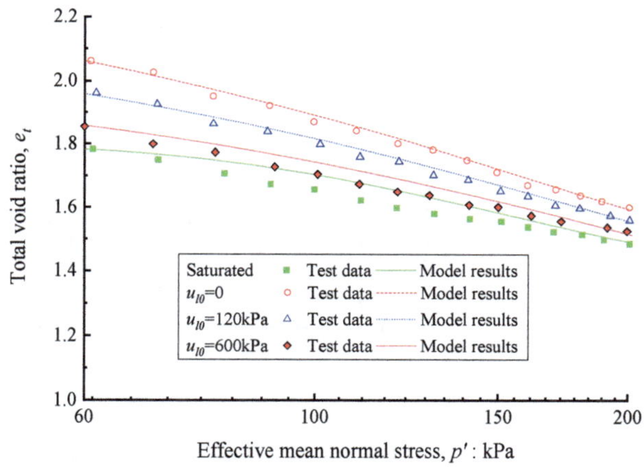

(b)

Fig. 5.9 Numerical simulations of consolidation tests: (**a**) water content; (**b**) total void ratio

soil sample (μ_{l0} = 120 kPa); when μ_{l0} = 600 kPa, the shear strength of the gas-bearing soil is less than that of the saturated soil sample; when μ_{l0} = 120 kPa, which is the same as the pore-water pressure of the saturated soil sample, its shear strength also basically remains unchanged, as shown in Fig. 5.10a. Similarly, the change in the pore-water pressure of gas-bearing fine-grained soil samples has a similar pattern to the change in shear strength. The magnitude of the increment of pore-water pressure ($\Delta\mu_l$) depends on the magnitude of the initial pore-water pressure, as shown in Fig. 5.10b. The change in the effective-stress path is shown in Fig. 5.10c. Since the gas-bearing specimen has a relatively large volume fraction of gas when μ_{l0} = 0, at the beginning of shearing, the undrained effective-stress path of the soil sample is closer to the drained path. However, when the initial pore-water pressure is high, the

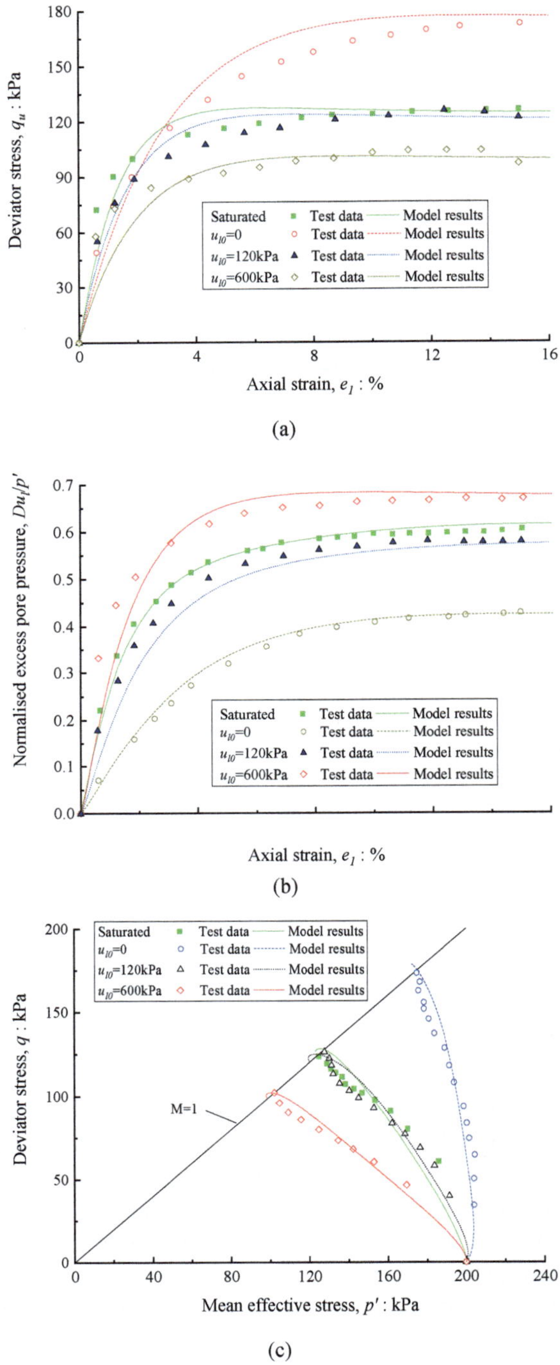

Fig. 5.10 Numerical simulations of undrained shear behavior of fine-grained gassy specimens at different initial pore pressures: (**a**) stress-strain relationship; (**b**) development of excess pore water pressure; (**c**) effective stress paths

undrained effective-stress path of the gas-bearing specimen has a greater contraction response than that of the saturated specimen. It can be seen that the simulation results are in agreement with the experimental data, indicating that the model can well reflect the undrained shear behavior of gas-bearing fine-grained soils under different initial pore-water pressures.

5.4 Discussion on the Temperature Characteristics of Fine-Grained Gassy Soils

Currently, there are no relevant literatures on the research of the temperature characteristics of gas-bearing soils. Since the occurrence conditions of gas-bearing soils will be affected by temperature, and due to the thermal expansion and contraction of the internal gas, the change in temperature will have a great impact on the engineering mechanical properties of gas-bearing soils. Therefore, the temperature characteristics of gas-bearing soils are a research direction worthy of exploration by scholars. The gas-bearing fine-grained soil model proposed in this chapter takes into account the influence of temperature changes. Due to the lack of comparison with relevant experiments, only qualitative analysis can be carried out. The experimental and theoretical research on the temperature characteristics of gas-bearing soils will also be the research work we will carry out in the future.

5.4.1 Influence of Temperature on Consolidation and Compression Characteristics

For normally-consolidated saturated soil, heating will cause thermal shrinkage, that is, thermal consolidation occurs. After heating, due to the thermal consolidation effect, the void ratio decreases, and the soil becomes denser. Its consolidation and compression curve will eventually tend to the consolidation and compression curve at normal temperature, that is, the consolidation and compression curve is independent of temperature [24], as shown in Fig. 5.11a. This means that the influence of temperature on the consolidation characteristics of gas-bearing soil samples mainly depends on the degree of influence of temperature on the gas phase. For gas-bearing soils, at normal temperature, the slope of the compression curve expressed by the total void ratio decreases with the increase of the gas content (gas volume), as shown in Figs. 5.4c and 5.9b. Under the same confining pressure, for a soil sample with a certain amount of gas in moles, when the temperature is high, according to the ideal gas state equation (Eq. (5.3)) and the relationship between gas pressure and confining pressure (Eq. (5.6)), the volume of the internal gas will increase. As known from the above-mentioned content, the increase of gas content will increase the compressibility of gas-bearing soil. Therefore, it can be concluded that heating will increase the compressibility of gas-bearing soil, as shown in Fig. 5.11b.

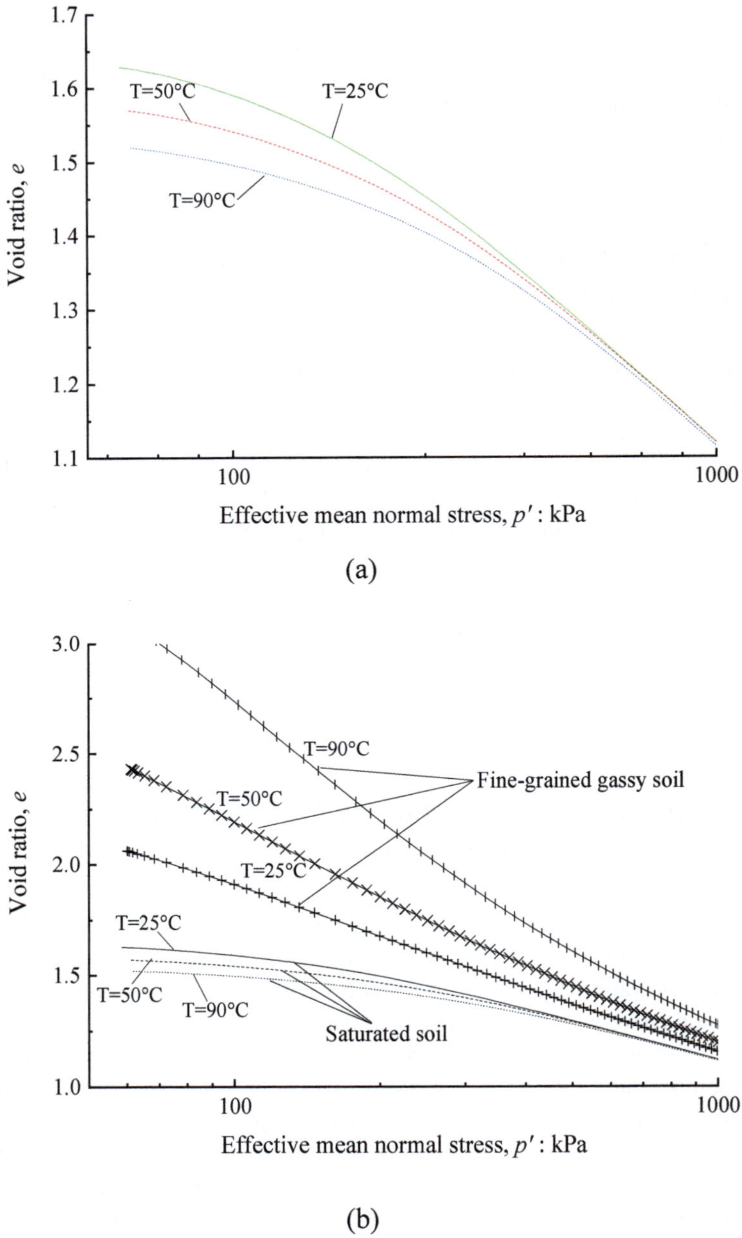

Fig. 5.11 Simulation of temperature effect on consolidation curves: (**a**) saturated soil; (**b**) fine-grained gassy soil

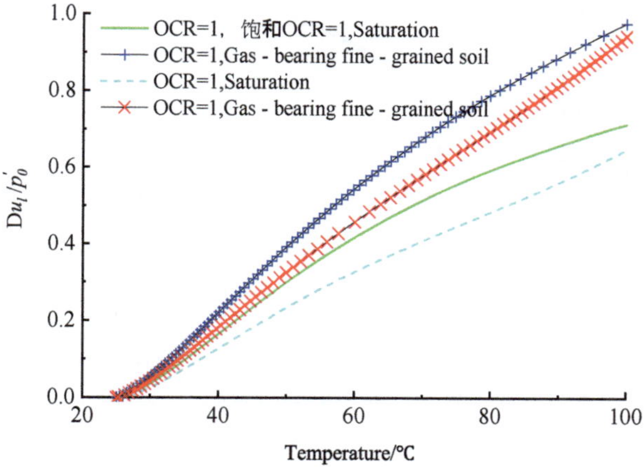

Fig. 5.12 Simulation of normalized thermally induced pore water pressures of saturated soil and fine-grained gassy soil with different overconsolidation ratios under undrained condition

In addition, under undrained conditions, due to the difference in the thermal expansion coefficients of the soil skeleton and pore water, heating will lead to an increase in the pore-water pressure of saturated soil [26]. Moreover, with the increase of the overconsolidation ratio (OCR), the increase in amplitude of the pore-water pressure shows a decreasing trend [28], as shown in Fig. 5.12. For isotropically loaded gas-bearing fine-grained soils, the presence of gas will increase the difference in the thermal expansion coefficients between the soil skeleton and the pore fluid, which will further lead to an increase in the pore-water pressure during undrained heating, as shown in Fig. 5.12.

Under drained conditions, for normally-consolidated saturated soil, heating leads to thermal shrinkage, also known as thermal consolidation. For overconsolidated saturated soil, heating may cause thermal expansion, and the larger the overconsolidation ratio, the more pronounced this phenomenon becomes [30–33]. In addition, for overconsolidated soil, as the temperature increases, the thermal strain may evolve from thermal expansion to thermal shrinkage, as shown in Fig. 5.13. Generally, the thermal expansion coefficient β_T of the soil skeleton varies with the OCR and temperature T. Usually, the relationship between β_T, OCR, and T is established to reflect the evolution law of thermal strain under different OCR values [29, 34], as in Eq. (4.35).

During the drained heating process, the expansion of gas will "squeeze out" more pore water. In other words, the expansion of gas will cause additional plastic volumetric strain in the saturated soil matrix (Eq. (5.14)). Therefore, for normally-consolidated soil bodies, the degree of thermal shrinkage of gas-bearing fine-grained soil is greater than that of saturated soil bodies, while for overconsolidated soil bodies, the degree of thermal expansion of gas-bearing fine-grained soil is less than that of saturated soil bodies, as shown in Fig. 5.13.

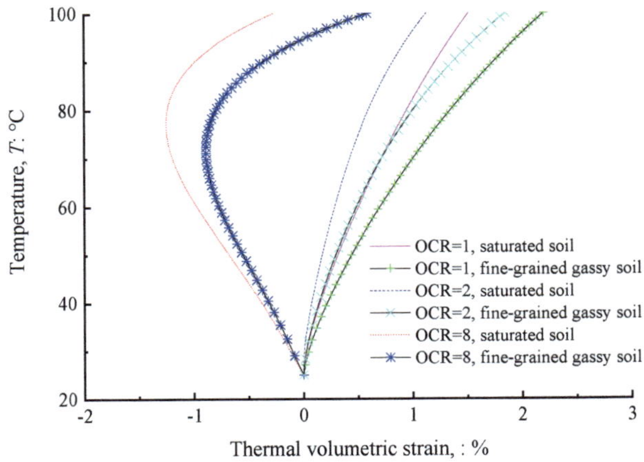

Fig. 5.13 Simulation of thermal volumetric strain of saturated soil and fine-grained gassy soil with different overconsolidation ratios under drained condition

5.4.2 Influence of Temperature on Undrained Shear Characteristics

The influence of temperature on the undrained shear strength of saturated soil is related to the state of the soil mass. According to relevant experimental research literatures [4, 23, 27], for normally-consolidated saturated soil, as the temperature increases, the undrained shear strength increases to a certain extent, as shown in Fig. 5.14a. For gas-bearing fine-grained soils, as known from the above, with the increase of gas content, the shear strength may increase or decrease, depending on the magnitude of the initial effective-like stress and confining pressure. When the temperature rises, the gas volume increases. When in conditions favorable for increasing the shear strength, the shear strength of the gas-bearing soil will further increase. Conversely, when in conditions unfavorable for increasing the shear strength, the increase in gas content will reduce the shear strength. Also, because the increase in temperature can increase the shear strength to a certain extent, under the combined influence of the two, it is not easy to determine whether the shear strength increases or decreases, and it is necessary to determine which factor has a more significant influence according to the test results. Figure 5.14b shows the simulation results of the influence of temperature on the undrained shear characteristics of gas-bearing fine-grained soils. Since the gas-bearing fine-grained soil is initially in conditions favorable for increasing the strength, when the temperature rises, the increase in amplitude of the shear strength of the gas-bearing fine-grained soil is greater than that of the saturated soil.

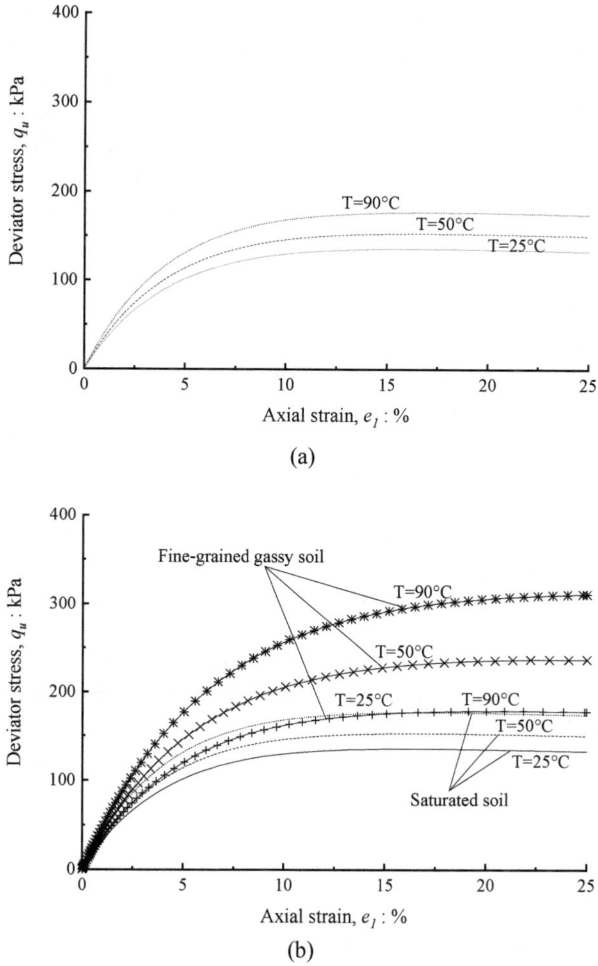

Fig. 5.14 Simulation of temperature effect on undrained shear strength: (**a**) saturated soil; (**b**) fine-grained gassy soil under the initial condition $u_l = 0$, $\sigma_3 = 200$ kPa

5.5 Summary

Gas-bearing fine-grained soil is a special type of unsaturated soil. The internal gas is in a closed state, and the gas pores are much larger than the average particle size. It exists in the form of discrete air bubbles in the completely saturated soil matrix and mostly appears in marine or lake sedimentary soils. Based on the concept of the large-air-bubble model proposed by Wheeler [11], gas-bearing fine-grained soil is regarded as composed of large air bubbles and the surrounding saturated soil matrix. On the basis of the thermomechanical model of saturated soil proposed in Chap. 4, combined with the state equation of an ideal gas and considering the influence of gas on the plastic deformation of the soil skeleton, a thermomechanical model of gas-bearing fine-grained soil is presented in this chapter. In the model, the gas pressure is controlled by the total stress and pore-water pressure and has a linear relationship with the total stress, while the saturated soil matrix part is controlled by the effective-like stress. Through the comparison between the simulation results and the test results, the ability of the model to describe the consolidation, compression, and undrained shear characteristics of gas-bearing fine-grained soil is verified.

In addition, the temperature characteristics of gas-bearing fine-grained soil are discussed, and the proposed model is used to analyze and simulate the temperature responses of gas-bearing fine-grained soil under different drainage conditions and overconsolidation ratios. Under drained conditions, an increase in temperature will increase the compressibility and thermal shrinkage of gas-bearing fine-grained soil with different overconsolidation ratios. Under undrained conditions, an increase in temperature will also lead to an increase in the excess pore-water pressure of gas-bearing fine-grained soil with different overconsolidation ratios. The influence of temperature on the undrained shear characteristics shows that when in conditions favorable for increasing the shear strength, an increase in temperature will further increase the shear strength. When in conditions unfavorable for increasing the shear strength, the increase or decrease of the shear strength depends on the difference in the influence degrees of temperature and gas volume fraction. If the influence degree of temperature is large, the shear strength increases; otherwise, it decreases. The model presented in this chapter can reflect the thermo-mechanical coupling characteristics of gas-bearing fine-grained soil.

References

1. Floodgate GD, Judd AG (1992) The origin of shallow gas. Cont Shelf Res 12(10):1145–1156
2. Hong Y, Wang LZ, Charles WWN, Yang B (2017) Effect of initial pore pressure on undrained shear behavior of fine-grained gassy soil. Can Geotech J 54(11):1592–1600
3. Fleischer P, Orsi T, Richardson M, Anderson A (2001) Distribution of free gas in marine sediments: a global overview. Geo-Mar Lett 21(2):103–122
4. Abuel-Naga HM, Bergado DT, Lim BF (2007) Effect of temperature on shear strength and yielding behavior of soft Bangkok clay. Soils Found 47(3):423–436
5. Wheeler SJ, Sham WK, Thomas SD (1990) Gas pressure in unsaturated offshore soil. Can Geotech J 27(1):79–89
6. Liu L, De Kock T, Wilkinson J, Cnudde V, Xiao SB, Buchmann C, Uteau D, Peth S, Lorke A (2018) Methane bubble growth and migration in aquatic sediments observed by X-ray μCT. Environ Sci Technol 52:2007–2015
7. Mabrouk A, Rowe RK (2011) Effect of gassy sand lenses on a deep excavation in a clayey soil. Eng Geol 122:292–302
8. Sobkowicz JC, Morgenstern NR (1984) The undrained equilibrium behaviour of gassy sediments. Can Geotech J 21(3):439–448
9. Grozic JLH, Imam SMR, Robertson PK, Morgenstern NR (2005) Constitutive modeling of gassy sand behaviour. Can Geotech J 42:812–829
10. Amaratunga A, Grozie JL (2009) On the undrained unloading behaviour of gassy sands. Can Geotech J 46(11):1267–1276

11. Wheeler SJ (1988a) A conceptual model for soils containing large gas bubbles. Géotechnique 38(3):389–397
12. Sills GC, Wheeler SJ, Thomas SD, Gardner TN (1991) Behavior of offshore soil containing gas-bubbles. Géotechnique 41(2):227–241
13. Sultan N, De Gennaro V, Puech A (2012) Mechanical behavior of gas-charged marine plastic sediments. Géotechnique 62(9):751–766
14. Sills GC, Gonzalez R (2001) Consolidation of naturally gassy soft soil. Géotechnique 51(7):629–639
15. Wheeler SJ (1986) Stress-strata behaviour of soils containing gas bubbles. Oxford University, Oxford
16. Wheeler SJ (1988b) The undrained shear strength of soils containing large gas bubbles. Geotech 38(3):399–413
17. Thomas SD (1987) The consolidation behavior of gassy soil. Oxford University, Oxford
18. Wheeler SJ, Gardner TN (1989) Elastic-moduli of soils containing large gas-bubbles. Géotechnique 39(2):333–342
19. Sultan N, Garziglia S (2014) Mechanical behavior of gas-charged fine sediments: model formulation and calibration. Géotechnique 64(11):851–864
20. Timoshenko SP, Goodier JN (1970) Theory of elasticity, 3rd edn. McGraw-Hill, New York
21. Pietruszczak S, Pande GN (1996) Constitutive relations for partially saturated soils containing gas inclusions. J Geotech Eng 122(1):50–59
22. Grozic JLH, Nadim F, Kvalstad TJ (2005) On the undrained shear strength of gassy clays. Comput Geotech 32(7):483–490
23. Cekerevac C, Laloui L (2004) Experimental study of thermal effects on the mechanical behaviour of a clay. Int J Numer Anal Methods Geomech 28:209–228
24. Yamamoto S, Alcauskas JB, Crozier TE (1976) Solubility of methane in distilled water and seawater. J Chem Eng Data 21(1):78–80
25. Bishop AW, Alpan I, Blight GE, Donald IB (1960) Factors controlling the strength of partly saturated cohesive soils. In: Research conference on shear strength of cohesive soils. American Society of Civil Engineers, New York, pp 503–532
26. Burghignoli A, Desideri A, Miliziano S (2000) A laboratory study on the thermomechanical behaviour of clayey soils. Can Geotech J 37(4):764–780
27. Kuntiwattanakul P, Towhata I, Ohishi K, Seko I (1995) Temperature effects on undrained shear characteristics of clay. Soils Found 35(1):147–162
28. Graham J, Tanaka N, Crilly T, Alfaro M (2001) Modified cam-clay modelling of temperature effects in clays. Can Geotech J 38(3):608–621
29. Laloui L, Cekerevac C (2008) Numerical simulation of the non-isothermal mechanical behaviour of soils. Comput Geotech 35:729–745
30. Abuel-Naga H, Bergado D, Bouazza A, Ramana G (2007) Volume change behaviour of saturated clays under drained heating conditions: experimental results and constitutive modeling. Can Geotech J 44(8):942–956
31. Bai B, Shi XY (2017) Experimental study on the consolidation of saturated silty clay subjected to cyclic thermal loading. Geomech Eng 12(4):707–721
32. Demars KR, Charles RD (1982) Soil volume change induced by temperature cycling. Can Geotech J 19(2):189–194
33. Sultan N, Delage P, Cui YJ (2002) Temperature effects on the volume change behaviour of boom clay. Eng Geol 64(2–3):135–145
34. Cui YJ, Sultan N, Delage P (2000) A thermomechanical model for saturated clays. Can Geotech J 37(3):607–620
35. Finno RJ, Zhang Y, Buscarnera G (2017) Experimental validation of Terzaghi's effective stress principle for gassy sand. J Geotech Geoenviron Eng 143(12):04017092

Chapter 6
Model Description of Mechanical Properties and Temperature Effects in Gassy Sand

6.1 Introduction

As a granular material, the mechanical properties of soil are very complex, especially for sandy soils, which are directly related to their physical state, and are characterized by the shear contraction of loose sand and the shear expansion of dense sand [1]. In addition, the loose and dense state of sandy soil is not only related to its relative density, but also depends on the amount of peripheral pressure exerted during the shear process. For sandy soils with the same relative density, shear swelling may occur at lower perimeter pressures while shear shrinkage may occur at higher perimeter pressures [1]. Saturated loose sands can even reach a state of static liquefaction during shear, which is a problem encountered from time to time in practical engineering. The development of an intrinsic model that can describe the complex mechanical properties of sandy soils is of high engineering and academic value.

In marine environments, such as river deltas, and deep-sea sediments, a special type of sandy soil, generally known as gas-bearing sand (gassy sand), often occurs. In addition, sandy soils are the main environment for shallow gas. Some of the structural characteristics of gassy sandy soils have been summarized in the introduction, i.e., small closed gas bubbles are free between the pores of the soil skeleton. Due to the compressibility of the gas, the presence of gas bubbles will have a certain impact on the engineering properties such as shear strength and bulk change characteristics of the soil, and even cause engineering accidents [2–5].

For the experimental study of aerated sandy soils, some scholars [6–12] demonstrated that the presence of a small number of gas bubbles in sandy soils can significantly affect the stress-strain behavior of loose sandy soils through many different testing techniques, including cyclic triaxial tests and torsional shear tests, and concluded that the advantages and disadvantages of the presence of gases are dependent on the loading paths and drainage conditions [3]. Vega-Posada et al. [13] investigated the triaxial compression and shear characteristics of gas-containing sandy

G. Yang, *Multi-field Coupled Constitutive Model for Soils Based on Granular Thermodynamics Theory*, https://doi.org/10.1007/978-981-95-0249-3_6

soils under drained and undrained conditions, and the test results showed that under undrained conditions, gas-containing sandy soils are more easily compressed than saturated sandy soils and have difficulty in generating superstatic pore water pressure. Under the drained condition, the shear strength of the aerated sandy soil was basically the same as that of the saturated sandy soil. He and Chu [14, 15] investigated the liquefaction behavior of aerated sandy soil under static monotonic loading condition by using undrained triaxial compression test and tensile test. The test results show that the saturation of loose sandy soil is reduced from 100% to in the range of 95% to 88% under undrained conditions, and the undrained shear strength can be increased by more than two times. The reduction of saturation also leads to a change in the stress-strain behavior from strain softening to strain hardening for sandy soils with a relative density of 10%. Experimentally investigated the influence of gas permeability, water-holding, stress path and gas release on the deformation properties and strength characteristics of gas-containing sandy soils, and gave countermeasures to prevent and control shallow gas hazards.

In the study of theoretical modeling of gas-containing sandy soils, it is generally assumed that the gas phase exists in the form of small bubbles within the pores of the soil skeleton, the surface tension of the bubbles is neglected, and the air pressure is equal to the sum of the external atmospheric pressure and the pore water pressure; the liquid phase is continuous and incompressible, and the gas phase is discontinuous and compressible; and the gas phase follows the Boyle's law and Henry's law. Based on the above assumptions, scholars often establish theoretical models for gas-containing sandy soils by taking into account the effect of gases (increased fluid compressibility) on the basis of some existing models. Pietruszczak and Pande [16] gave a mathematical model that can describe the undrained shear properties of sandy soils containing small bubbles. Grozic et al. [17] gave a mathematical model that can describe the undrained shear properties of sandy soils containing small bubbles on the basis of existing models by taking into account pore space model. An intrinsic model that can reflect the mechanical properties of air-containing sandy soils was developed by considering the compression and dissolution properties of pore gas. Liu et al. [18] developed a mathematical model that can describe the consolidation properties of geotechnical media containing organic matter and small bubbles, and regarded the consolidation of the material as a coupled process of pore water drainage and the change in the volume of gas bubbles and organic matter. Lü et al. [19] developed a theoretical model that can predict the static liquefaction phenomenon of air-bearing sandy soils under triaxial undrained conditions by improving the partial hardening plasticity model.

Despite this particular mechanical property of sandy soils, however, no matter what physical state they are in initially, they will eventually reach a critical state where the body strain is zero [20]. Some previous sand-soil constitutive models, which regard sand-soil in different physical states as different materials and use different model parameters for simulation, cannot truly and completely describe the mechanical properties of sand-soil during the loading process [21, 22]. In order to overcome the above shortcomings, many scholars have established the intrinsic model of sandy soil by introducing state parameters on the basis of critical state theory [23–30]. The state parameter is jointly determined by the two factors of relative density and circumferential pressure, which can reflect the loose and dense state of the material, so that the established constitutive model can consider the mechanical behavior due to the change of state during the loading process. In addition, some scholars believe that the critical state of sandy soils is more difficult to be obtained in tests, and that using the critical state line as the reference line of the state parameter may lead to the inability to accurately characterize the deformation properties of sandy soils, and they take the phase change state as a reference point to establish a phase change state-based of the boundary surface intrinsic model.

Similarly, most of the sandy soils modeling is based on classical elastoplasticity theory or on thermodynamic principles by defining the free energy function and the dissipation potential function to model the elastoplasticity or superplasticity eigenstructure. Based on the thermodynamic model based on the thermodynamic theory of particulate matter presented in Chap. 2, a thermodynamic constitutive model describing the shear swelling properties of saturated sandy soils is developed by combining the shear swelling equation with the introduction of state parameters. The model introduces the concept of particle temperature to describe the irreversible deformation within the sandy soil, and links the dissipation mechanism of the saturated sandy soil to the macroscopic mechanical behavior through the migration coefficient and energy density function. The model can describe both the effect of saturated sandy soils on the strength and deformation characteristics of the soil due to the changes in relative density and effective confining pressure during shear. For air-containing sandy soils, it is considered that the presence of air bubbles only increases the compressibility of the pore fluid and does not have an effect on the deformation of the soil skeleton. Based on the thermodynamic model of saturated sandy soils, and in combination with the relationship between pore gas pressure and change in the volume of pore gas established by Hilf [37], a thermodynamic ontological model that can characterize the undrained shear properties of air-containing sandy soils has been developed. The validity of the model is verified based on the comparison of simulations with results of isotropic compression, triaxial undrained, and drained shear tests.

6.2 Thermodynamic Principal Modeling of Saturated Sand

The theory of thermal dynamics of particulate matter is more applicable to the study of sandy soils due to the fact that the effect of cohesion is not taken into account. In the framework of the theoretical model for unsaturated soils presented in Chap. 2, for the case of saturated sandy soils, the effect of temperature is not taken into account, combined with the shear expansion equation considering the parameters of the state, and through the modification of some of the constitutive relations, so that it is possible to develop a thermodynamic constitutive model of saturated sandy soils that describes the shear properties of saturated sandy soils.

6.2.1 Expressions for Stress

The equilibrium thermodynamic full differential form of a sandy soil can be expressed as:

$$d_t w = \pi_{ij} d_t \varepsilon_{ij}^e + T_g d_t S_g + T d_t S + \sum_{\alpha=S,L} \left(\mu_{c\alpha} d_t \rho^\alpha + v_t^\alpha d_t m_t^\alpha \right) \tag{6.1}$$

This leads to some thermodynamic equations such as $\pi_{ij} = \partial w / \partial \varepsilon_{ij}^e$. The work done by the stress on the elastic deformation of a saturated sandy soil during stress will be transformed into the elastic potential energy density of the soil body WE, and thus $\pi_{ij} = \partial w_e / \partial \varepsilon_{ij}^e$ can be obtained.

Zhang and Cheng [31] corrected the elastic potential energy density function (Eq. (2.62)) given by Jiang and Liu [32] for a clayey soil material with water in the pores as:

$$w_e = B\left(\varepsilon_v^e + c\right)^\beta \left[\left(\varepsilon_v^e\right)^2 + \zeta \left(\varepsilon_s^e\right)^2 \right] \tag{6.2}$$

where ζ is the material parameter; and analyzed that the case of geotechnical materials is more satisfied when β is taken as 1.5.

In the study in Chap. 2, for unsaturated clayey soils, the elastic potential energy density function used is Eq. (2.64), which follows the basic form given by Jiang and Liu [32], which is equivalent to β taken to be 0.5. Furthermore, since the elastic potential energy density function can be determined based on the influence of mean effective stress on the elastic bulk modulus and elastic shear modulus. In the case of sandy soil, during the shear process, liquefaction may occur, i.e., the effective stress is 0, and the elastomeric strain ε_v^e tends to be 0. For this phenomenon, according to the expression of the average effective stress (Eq. (6.8) below), β is more appropriate to be taken as 1.5. Therefore, for the case of saturated sandy soil, without considering the effect of cohesion, Eq. (6.2) can be modified as:

$$w_e = B\left(\varepsilon_v^e\right)^{1.5} \left[\left(\varepsilon_v^e\right)^2 + \zeta \left(\varepsilon_s^e\right)^2 \right] \tag{6.3}$$

This leads to an expression for π_{ij}:

$$\pi_{ij} = \frac{\partial w_e}{\partial \varepsilon_{ij}^e} = B\left[3.5\left(\varepsilon_v^e\right)^{2.5} \delta_{ij} + 1.5\zeta \left(\varepsilon_v^e\right)^{0.5} \left(\varepsilon_s^e\right)^2 \delta_{ij} + 2\zeta \left(\varepsilon_v^e\right)^{1.5} e_{ij}^e \right] \tag{6.4}$$

Similarly, the elastic bulk modulus K_e and elastic shear modulus G_e are defined as:

$$K_e = B\left[3.5\left(\varepsilon_v^e\right)^{1.5} + 1.5\zeta \left(\varepsilon_v^e\right)^{-0.5} \left(\varepsilon_s^e\right)^2 \right] \tag{6.5}$$

$$G_e = B\zeta \left(\varepsilon_v^e\right)^{1.5} \tag{6.6}$$

π_{ij} can in turn be expressed in a form similar to Cauchy stresses:

$$\pi_{ij} = K_e \varepsilon_v^e \delta_{ij} + 2G_e e_{ij}^e \tag{6.7}$$

In addition, the average effective mean stress and shear stress can be expressed respectively as:

$$p' = \frac{\pi_{kk}}{3} = B\left[3.5\left(\varepsilon_v^e\right)^{2.5} + 1.5\zeta\left(\varepsilon_v^e\right)^{0.5}\left(\varepsilon_s^e\right)^2\right] \tag{6.8}$$

$$q = \sqrt{\frac{3}{2}s_{ij}s_{ji}} = \sqrt{6}\zeta B\left(\varepsilon_v^e\right)^{1.5}\varepsilon_s^e \tag{6.9}$$

From the isotropic consolidation compression curves of saturated sandy soils, it can be seen that the pore ratio e is approximately linearly related to lnp':

$$\Gamma = e + \lambda \ln p' \tag{6.10}$$

where λ is the slope of the compression curve; Γ is the special pore ratio atp' = 1 kPa.

Isotropic solidified compression, $\varepsilon_s^e = 0$. Taking logarithms on both sides of Eqs. (6.8) yields, lnp' = ln B + ln A, where $A = 3.5\left(\varepsilon_v^e\right)^{2.5}$. Assuming that B can be expressed in the form of B_0 exp Ξ, we have:

$$\ln p' = \Xi + \ln B_0 + \ln A \tag{6.11}$$

Comparing Eq. (6.11) with Eq. (6.10), the form of B can also be approximated, $B = B_0 \exp\left(-\frac{e}{\lambda}\right)$. Since B is a parameter that describes the degree of softness or hardness of a material, the denser the soil, the greater the value of B. The expression for B obtained above can reflect this property.

6.2.2 Expression of Strain

Irreversible deformation of soil bodies is due to inelastic interactions between particles. The introduction of particle temperature describes the strength of the inter-particle interaction and relates the dissipation mechanism of the soil body to the macroscopic deformation, thus determining the magnitude of the irreversible deformation of the soil body. The inelastic interaction between particles produces dissipation, and the effective stress π_{ij} is the dissipative force that causes an increase in entropy, and the corresponding dissipative flow is the irreversible deformation rate, i.e., the plastic deformation rate $d_t\varepsilon_{ij}^p$. Since there is a linear relationship between the dissipative flow and the dissipative force, there are:

$$d_t\varepsilon_{ij}^p = \eta_{ijkl}\tau_{kl} \tag{6.12}$$

where η_{ijkl} is the migration coefficient, which is related to the particle temperature, and similarly, under isotropic conditions, its form can be defined as:

$$\eta_{ijkl} = (T_g)^\alpha \frac{\eta_s}{2G_e} \delta_{il}\delta_{jk} + (T_g)^\alpha \left(\frac{\eta_v}{3K_e} - \frac{\eta_s}{6G_e} \right) \delta_{ij}\delta_{lk} \tag{6.13}$$

where η_v and η_s are the corresponding migration coefficients; the index α is a parameter reflecting the rate correlation [31], which is rate-independent when it is taken as 0.5, and α is taken as 0.5 in this model.

Substituting Eqs. (6.8) and (6.13) into Eq. (6.12), the expression of plastic strain rate can be obtained as:

$$d_t\varepsilon_{ij}^p = \eta_v (T_g)^{0.5} \varepsilon_v^e \delta_{ij} + \eta_s (T_g)^{0.5} e_{ij}^e \tag{6.14}$$

Let $c_1 = \eta_v/\eta_s$ $T_{gg} = \eta_s^2 T_g$, then the elastic strain rate $d_t\varepsilon_v^e$ and the elastic shear strain rate $d_t\varepsilon_s^e$ of the soil can be expressed respectively as:

$$d_t\varepsilon_v^e = d_t\varepsilon_v - 3c_1 (T_{gg})^{0.5} \varepsilon_v^e \tag{6.15}$$

$$d_t\varepsilon_s^e = d_t\varepsilon_s - (T_{gg})^{0.5} \varepsilon_s^e \tag{6.16}$$

6.2.3 State Parameter

According to the theory of critical state geotechnics [20], when the critical state is reached, the bulk strain no longer changes, and at this point there is a unique relationship between the pore ratio and the effective stress acting on the soil. For clayey soils, when the critical state is reached, the critical pore ratio e_c is linearly related to the effective stress p' in the e_c-lnp' plane [33], and a similar pattern is found for sandy soils [36].

$$e_c = \Gamma - \lambda \ln(p') \tag{6.17}$$

where λ is the slope of the curve formed in the e_c-lnp' plane, and Γ is the critical porosity ratio at $p' = 1$ kPa.

Others have argued that the critical state line of sandy soil materials is more suitable to be described by a power function, Li and Dafalias [29] expressed the critical state line of sandy soil as:

$$e_c = e_r - \lambda_c (p'/p_a)^\varsigma \tag{6.18}$$

where λ_c is the slope of the curve in the$e - (p'/p_a)^{\varsigma}$plane, as shown in Fig. 6.1; e_r, ς are the parameters of the curve; p_a is the atmospheric pressure. Equation (6.18) is used in this chapter as an expression for the critical state line of sandy soil.

Been and Jefferies [24] introduced a state parameter to describe the loose and dense state of sandy soil, defined as $\psi = e - e_c$, which is the difference between the current pore ratio e and the critical state pore ratio e_c with the same effective stress as e, as shown in Fig. 6.1. When $\psi > 0$, the sandy soil is in a loose state, and vice versa. The definition of state parameter includes both relative density and pressure, which can reflect the state change of sandy soil during the loading process. Some scholars [30] also define the state parameter as $\psi = e - e_c$, and the significance of the two is equivalent.

6.2.4 Shear Expansion Equation

Shear expansion is one of the basic properties of geotechnical materials, and a reasonable description of the shear expansion equation plays a decisive role in establishing the intrinsic model of sandy soils. The ratio of the plastic body strain increment to the plastic shear strain increment is generally defined as the shear expansion ratio d, $d = d_t\varepsilon_v^p / d_t\varepsilon_s^p$. Many different forms of the shear expansion equation exist in the existing literature [28–30], Li and Dafalias [29] introduced the state parameters into the shear expansion equation and established a shear expansion expression related to the state of sandy soil:

$$d = d_0\left(e^{m\psi} - \frac{\eta}{M}\right) = \frac{d_0}{M}\left(Me^{m\psi} - \eta\right) \tag{6.19}$$

where $\eta = q/p'$ is the stress ratio; $\psi = e - e_c$ is the state parameter; M is the critical state stress ratio; d_0, m and m are non-negative constants, which can be determined

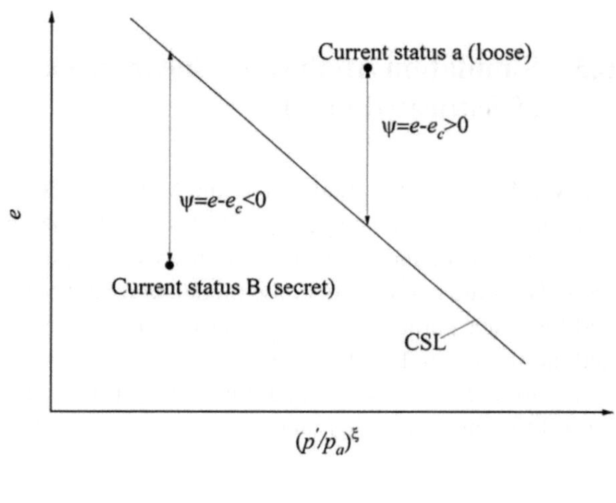

Fig. 6.1 Sketch of critical state line and state parameter

by test. When $m = 0$, $d_0 = M$, Eq. (6.19) degenerates into Cambridge shear expansion formula, $d = M - \eta$.

From Eqs. (6.15) and (6.16), $d_t\varepsilon_v^p = 3c_1\left(T_{gg}\right)^{0.5}\varepsilon_v^e$, $d_t\varepsilon_s^p = \left(T_{gg}\right)^{0.5}\varepsilon_s^e$. Combined with the shear expansion Eq. (6.19), the expression for c_1 is given by:

$$c_1 = \frac{d_0}{3M\varepsilon_v^e}\left(Me^{m\psi} - \eta\right)\varepsilon_s^e \tag{6.20}$$

6.2.5 Calibration of Model Parameters

1. The calibration of the parameters related to the shear expansion equation can be determined by reference to the literature [29], which gives the specific method of determination. M is the critical stress ratio; e_Γ, ς, λ_c can be fitted according to the experimental data of the critical state line; m can be determined according to the state of the phase transition at $d = 0$; and d_0 can be determined according to the $\varepsilon_v \sim \varepsilon_s$ curve.
2. Parameters related to the elastic potential energy density function, B_0 and λ can be determined by isotropic consolidation curve (Eq. (6.8)), and ζ can be determined by the stress ratio and the magnitude of shear stress when the critical state is reached.
3. The parameter c_1 related to the strain evolution can be determined according to Eq. (6.20); c_2 is related to the elastic shear strain at the critical state, and when the critical state is reached, it can be obtained from Eq. (2.18) and Eq. (6.16), $\left(\varepsilon_s^e\right)_{cr} = C_2^{0.5}$, and the expression can be obtained by combining with the stress ratio at the critical state (Eq. (6.8) and Eq. (6.9)); c_3 is related to the solidification compression curve, and it can be derived from the isotropic solidification compression test; c_4 can be calibrated by stress relaxation test as described in Chap. 3, Sect. 3.1.

6.3 Simulation Analysis of Shear Properties of Saturated Sand

The validity of the thermodynamic model for saturated sandy soils presented in this chapter was verified using a series of tests performed on Toyoura sand in the literature [35, 36], including isotropic compression tests, drained and undrained shear tests. The state parameters in the model were taken directly from the literature [29] and the other parameters were determined by the method described in Sect. 6.2.5 and the values are listed in Table 6.1.

Figure 6.2 shows the simulation results of the compression test of Toyoura sand with different pore ratios, and it can be seen that the slopes of the consolidation

Table 6.1 Model parameters

State parameters			
M	1.25	λ_c	0.019
ς	0.7	m	3.5
e_Γ	0.934	d_0	10
Thermodynamic model parameters			
c_3	90,000 s^2	c_4	800 kg/(m^3·s)
B_0	1.064 × 10^8 MPa	λ	0.45
ζ	15	G_s	2.64

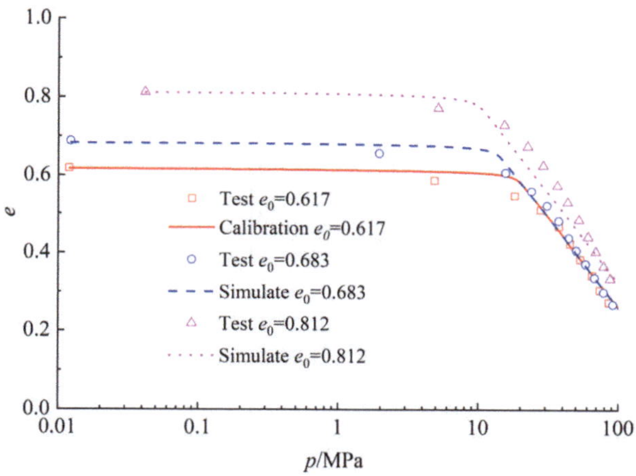

Fig. 6.2 Simulation of compression tests of Toyoura sand with different void ratios

compression curves are basically the same in the *e*-ln*p'* plane for sands with different densities, and the simulation results of the model can be in good agreement with the test results.

Figures 6.3, 6.4, and 6.5 show the simulation results of undrained shear tests for specimens with different pore ratios $e = 0.735/0.833/0.907$ at different initial perimeter pressures $\sigma_3 = 1/2/3$ MPa, respectively. It can be seen that when the sandy soil pore size is relatively small (Fig. 6.3), the specimens under different perimeter pressures show shear hardening and shear shrinkage followed by shear expansion. When the initial perimeter pressure is higher, the soil samples are relatively loose and the amount of shear shrinkage is also larger, while the subsequent shear swelling is relatively smaller. As the pore ratio increases (Fig. 6.4), the shear shrinkage of the specimen increases and the subsequent shear swelling decreases. When the perimeter pressure is large, essentially no shear swelling occurs and significant softening occurs. When the pore size is relatively large (Fig. 6.5), the specimens under different peripheral pressures show obvious softening phenomenon, and the specimens always show shear shrinkage. It can be seen that the model simulation results can better reflect the undrained shear characteristics of saturated sandy soil.

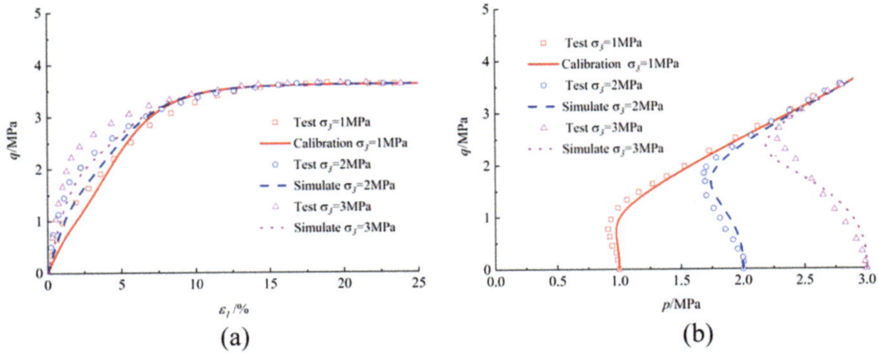

Fig. 6.3 Simulation of undrained triaxial tests for $e = 0.735$: (**a**) stress-strain curves; (**b**) effective stress path

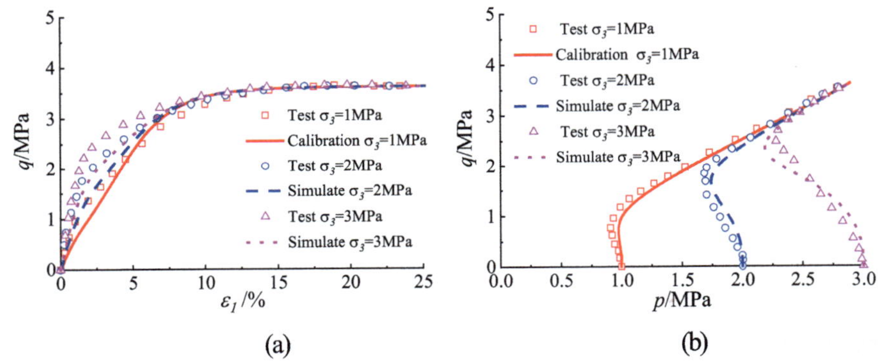

Fig. 6.4 Simulation of undrained triaxial tests for $e = 0.833$: (**a**) stress-strain curves; (**b**) effective stress path

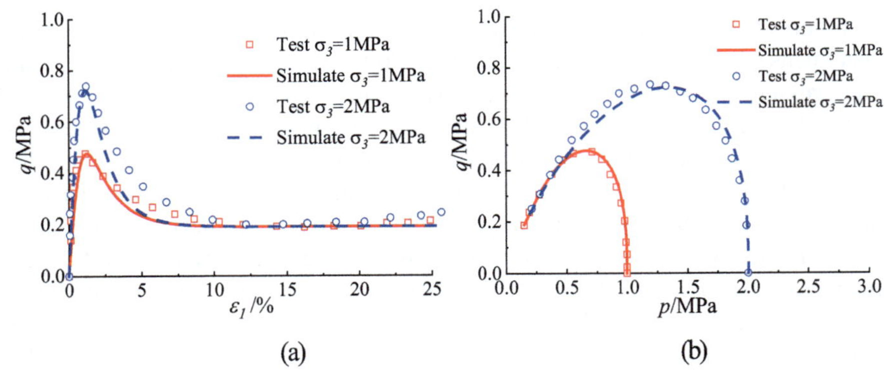

Fig. 6.5 Simulation of undrained triaxial tests for $e = 0.907$: (**a**) stress-strain curves; (**b**) effective stress path

Figures 6.6 and 6.7 show the simulation results of drainage shear tests at initial perimeter pressures of 0.1 MPa and 0.5 MPa for specimens with different pore ratios $e = 0.831/0.917/0.996$, respectively. When the pore size is relatively large, the specimens consistently show shear shrinkage and exhibit shear hardening, and when the pore size is relatively small ($e = 0.831$), the specimens first undergo shear shrinkage and then shear swelling, accompanied by a slight softening phenomenon. It can be seen that the overall simulation results are basically consistent with the test results, but there is a certain difference between the simulation results and the test results for when the porosity ratio is larger, and the rate of increase of the shear stress in the simulation calculation is larger, but the overall can reflect the trend of the volume change.

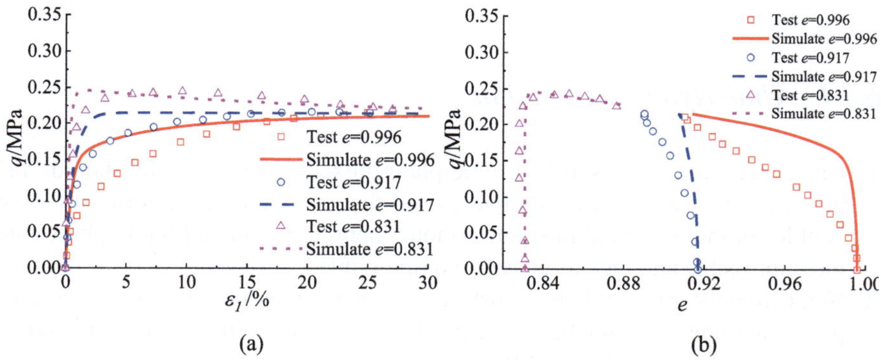

Fig. 6.6 Simulation of drained triaxial tests for $p_0' = 0.1$ MPa: (**a**) stress-strain curves; (**b**) void ratio versus deviator stress

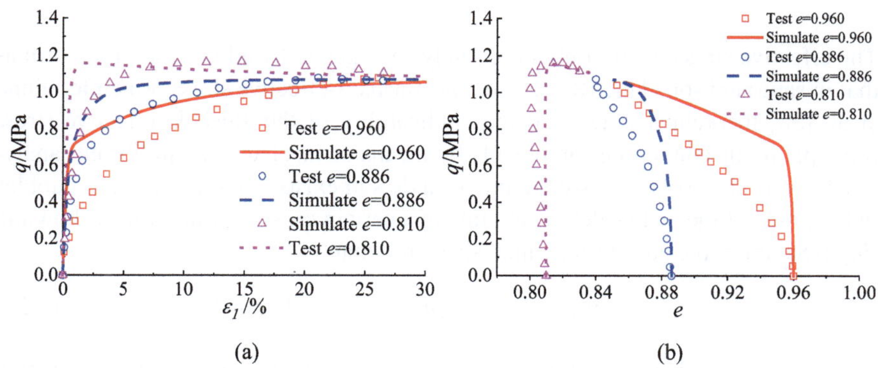

Fig. 6.7 Simulation of drained triaxial tests for $p_0' = 0.5$ MPa: (**a**) stress-strain curves; (**b**) void ratio versus deviator stress

6.4 Thermodynamic Modeling of Gassy Sand

Some scholars [17, 19] have developed theoretical models for gas-bearing sandy soils based on some existing models for saturated sandy soils, taking into account the effect of gases (increased compressibility of fluids). Similarly, based on the thermodynamic model of saturated sandy soils developed above, a thermodynamic ontological model of gas-bearing sandy soils was developed by incorporating the relationship between the pore gas pressure and the change in pore gas volume given by Hilf [37]. The ability of the model to describe the undrained shear properties of air-bearing sandy soils was verified by triaxial undrained shear tests conducted by Grozic et al. [4] on remodeled Ottawa sandy soils with different degrees of saturation.

6.4.1 Underlying Assumption

1. Gas-containing sandy soil is a three-phase mixed system of solid, liquid, and gas, with the liquid phase continuous and the gas phase in the form of closed bubbles in the pore fluid medium. Among them, the solid and liquid phases are incompressible and the gas phase is compressible.
2. Neglecting the effect of bubble surface tension, the pressure of the gas phase, the pore water pressure, and the pressure of the water-gas mixture fluid are the same.
3. Gases follow Boyle's law and Henry's law.

6.4.2 Effective Stresses of Gassy Sand

The effective stress in air-containing sandy soils generally adopts the same form as that in saturated soils [17, 18, 38], i.e., the Taishaki-effective stress principle. Finno et al. [38] theoretically deduced the applicability of the Taishaki-effective stress principle in air-containing sandy soils by means of mechanical equilibrium analysis. The gas phase of the gas-containing sandy soil exists in the form of gas bubbles in the pores of the soil skeleton, and taking a certain cross-section of it as shown in Fig. 6.8a, the force equilibrium analysis shows that:

$$\sigma_A = \sigma_z A_s + u_i A_1 + \sum u_{g,i} A_{g,i} - T_z \sum l_{g,i} \sin\theta_i \tag{6.21}$$

where σ is the total applied stress; A is the cross-sectional area; σ_s is the force acting on the soil particles; A_s is the area of the soil particles; u_i is the pressure of the pore liquid phase; $A_1 u_i$ is the area of the liquid phase; $u_{g,i} A_1$ is the pressure of the individual bubbles; $A_{g,i}$ is the cross-sectional area of the individual bubbles; T_s is the

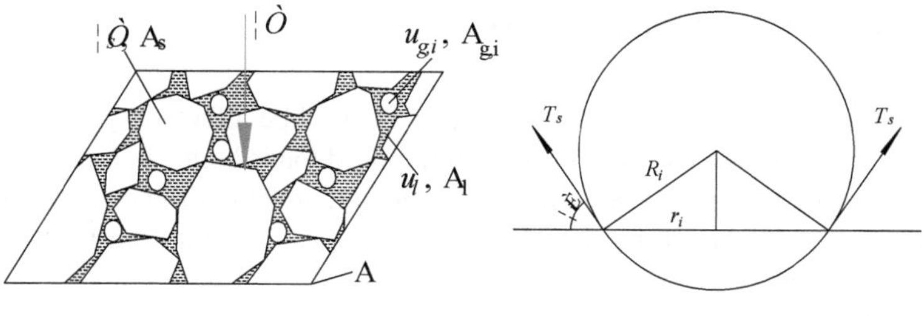

Fig. 6.8 (**a**) An ideal cut plane of gassy sand. (**b**) Radius of bubble cut and surface tension

surface tension; $l_{g,i}$ is the perimeter of the cross-section of the bubbles; and θ_i is the solid-liquid-phase contact angle, as shown in Fig. 6.8b.

Also, because, the relationship between pore water pressure and air pressure is

$$u_{g,i} = u_l + 2T_s / R_i \tag{6.22}$$

where R_i is the radius of curvature of the bubble.

The last two terms in Eq. (6.21) can be simplified as follows:

$$\begin{aligned} \Sigma u_{g,i}A_{g,i} - T_s \Sigma l_{g,i} \sin\theta_i &= \Sigma\left(u_l + \frac{2T_s}{R_i}\right)A_{g,i} - T_s \Sigma 2\pi r_i \sin\theta_i \\ &= u_l \Sigma A_{g,i} + T_s \Sigma\left(\frac{2\pi r_i^2}{R_i} - 2\pi r_i \sin\theta_i\right) = u_l A_g \end{aligned} \tag{6.23}$$

Substituting Eq. (6.23) into Eq. (6.21) yields:

$$\sigma_A = \sigma_z A_s + u_i A_1 + u_i A_g = \sigma_z A_s + u_i (A - A_s) \tag{6.24}$$

Dividing both sides simultaneously by the total area A yields:

$$\sigma = \frac{\sigma_s A_s}{A} + u_l\left(1 - \frac{A_s}{A}\right) \tag{6.25}$$

where $\sigma_s A_s/A$ is the average stress on the cross-sectional area A of the contact stress between soil particles, i.e., the effective stress σ' of the Taiwanese sand base. It is shown that the contact area A_s between soil particles is very small, so A_s/A in the second term on the right side of the equal sign of Eq. (6.25) is negligible, but in the first term, σ_s cannot be omitted because it is large. Therefore, Eq. (6.25) can be simplified to $\sigma = \sigma' + u_l$, which shows that the principle of effective stress in tai-sa base is still applicable to airy sandy soils.

6.4.3 Characterization of Pore Fluids

The mechanical properties of aerated sandy soils will be affected by a combination of pore water and pore gas pressures, with imposed loads being an important factor in the variation of pore fluid pressures. In particular, under undrained conditions, the applied stresses are borne by the soil skeleton, pore gas, and pore water, and the weight borne depends on their relative compressibility.

Similarly, the pore gas ratio e_g, pore water ratio e_l, and total pore ratio e are defined as:

$$e_g = \frac{V_g}{V_s} = (1 - s_r)e \tag{6.26a}$$

$$e_l = \frac{V_l}{V_s} = s_r e \tag{6.26b}$$

$$e = e_1 + e_g \tag{6.26c}$$

where V_g, V_l, and V_s are the volumes of the gas, liquid, and solid phases, respectively.

6.4.4 Boyle's Law and Henry's Law

Liquid and gas phases in pore fluids exist in three forms: liquid and gas completely miscible, liquid and gas completely immiscible, and liquid and gas partially miscible. In the case of mutually miscible fluids, there is interaction between liquid and gas, while in the case of immiscible fluids, there is no interaction between liquid and gas. The presence of free gas will increase the compressibility of the soil, and if the liquid and gas phases are miscible to some extent, the compressibility of the soil will be a function of the pore pressure, the gas compression coefficient, and the dissolution coefficient. In the case of gas-containing sandy soils, the gas bubbles are inside the pores and the effect of surface tension on the soil skeleton can be neglected. Based on this, the water-gas mixture can be regarded as a homogeneous medium, assuming that the gas is an ideal gas, which satisfies Boyle and Henry's law, and the pore gas pressure is equal to the pore water pressure plus the atmospheric pressure, according to the volume change of air-containing sandy soils after loading, as shown in Fig. 6.9, there are:

$$(p_{atm} + u_{l_0})(V_{g_0} + hV_{l_0}) = (p_{atm} + u_l)(V_g + hV_l) \tag{6.27}$$

Among them, p_{atm} is a standard atmospheric pressure, 101.325 kPa; h is the solubility coefficient (Henry constant), which can be taken as 0.02 for water gas fluid mixtures [39]; Vg and V_l are the volumes of pore gas and pore water, respectively.

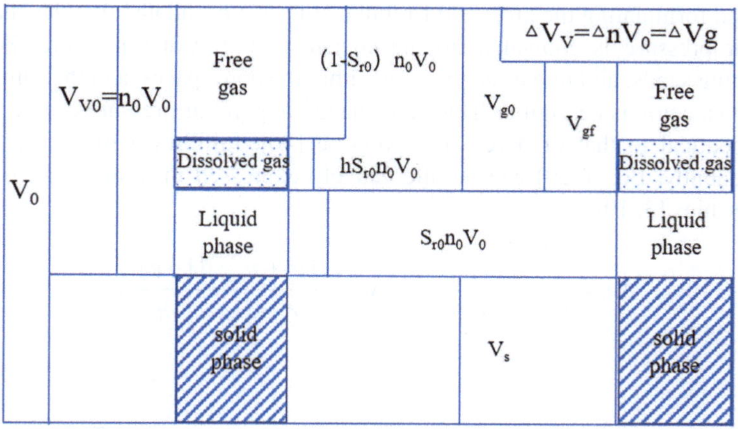

Fig. 6.9 Initial and final volume change of gassy sand

Under undrained conditions, pore gas and pore water are not allowed to be discharged, so any applied stress will cause an increase in pore pressure. For gas bearing sandy soil, due to the high compressibility of gas, the applied stress will also cause a change in volume. Hilf [37] established the relationship between pore gas pressure and pore gas volume change under undrained loading conditions based on Boyle's law and Henry's law:

$$\Delta u_g = \left(\frac{\Delta n}{(1 - s_{r0} + hs_{r0})n_0 - \Delta n} \right) u_{g0} \tag{6.28}$$

Among them, Δu_g is the change in air pressure, u_{g0} is the initial absolute air pressure, s_{r0} is the initial saturation, n_0 is the initial porosity, and Δn is the change in porosity.

Equation (6.28) describes the relationship between the pore gas pressure and the pore gas volume during the undrained loading process. By transforming Eq. (6.28), the volume change of the gas-containing sand during the loading and unloading processes can be obtained:

$$\frac{\Delta V}{V_0} = \left(\frac{\Delta u_g}{u_{g0} + \Delta u_g} \right) (1 - s_{r0} + hs_{r0}) n_0 \tag{6.29}$$

Among these parameters, $\frac{\varnothing V}{V_0}$ represents the relative change of the pore volume with respect to the initial volume.

Equation (6.29) characterizes the variation of volumetric strain resulting from gas compression. After each stress increment, both the saturation and the void ratio are updated. The newly calculated void ratio and saturation will then be incorporated into the subsequent calculation.

When formulating the aforesaid formula, Hilf [37] postulated that the change in pore-air pressure is equivalent to the change in pore-water pressure. For gas-containing sands, air bubbles are present within the pore spaces, and their impact on the soil skeleton is negligible. Thus, the change in pore-air pressure can be considered identical to that of pore-water pressure [40], fulfilling Hilf's assumptions. Consequently, Eq. (6.29) can be alternatively expressed in terms of incremental relationships [3, 19].

$$\Delta u_l = \frac{u_l + p_{atm}}{\left(1 - s_{r0} + hs_{r0}\right)n_0}\Delta\varepsilon_v = \frac{\left(u_l + p_{atm}\right)\left(1+e\right)}{e_g + \left(e - e_g\right)h}\Delta\varepsilon_v \tag{6.30}$$

6.5 Simulation Analysis of the Shear Characteristics of Gassy Sand

Grozic et al. [4] investigated the undrained shear characteristics of gas-containing sands through triaxial undrained shear tests on remolded Ottawa sands with different saturations. The test results indicate that gas-containing sands exhibit strain softening or strain hardening during undrained shearing, depending on the magnitudes of the initial saturation and the void ratio. When the initial void ratio of the gas-containing sand is relatively large, it generally shows strain softening and may even experience liquefaction. When the saturation is relatively small (<80%), the gas-containing sand generally exhibits strain hardening. The results of several representative specimens with different saturations were selected to verify the ability of the proposed gas-containing soil model to describe the undrained shear characteristics of gas-containing sands, as shown in Table 6.2. Among them, specimens 11/12 are saturated specimens, and the results of the saturated specimens are used to calibrate some model parameters. The others are specimens with different saturations and initial void ratios. The undrained shear rate is 0.375%/min, and the parameters required by the model are listed in Table 6.3.

Table 6.2 Sample description

Sample	Initial confining pressure(kPa)	Initial saturation	Void ratio	Maximum (kPa)		Minimum (kPa)		Final value (kPa)	
				p′	q	p′	q	p′	q
10	258.6	0.979	0.852	195.1	155.7	37.5	75.3		
11	266.3	1	0.896	197.9	121.9	10.5	32.9		
12	302.1	1	0.910	240.3	134.4	17.0	29.8		
16	317.1	0.816	0.876					472.9	713.6
21	296.3	0.802	0.897					484.7	799.6
25	271.7	0.912	0.852	226.1	117.3	9.6	0.0		

Table 6.3 Model parameters

Status parameters			
M	1.19	λ_c	0.0159
h	0.02	m	3.5
e_{c0}	0.92	d_0	50
Thermodynamic model parameters			
c_3	80,000 s^2	c_4	800 kg/(m^3·s)
B_0	2.128×10^{12} kPa	λ	0.25
ζ	7	G_s	2.67

The consolidation-compression tests of Ottawa sands in Ref. [41] were employed to calibrate some of the model parameters and validate the model's capacity to simulate the consolidation-compression characteristics. Figure 6.10 presents the simulation of the compression curves of saturated Ottawa sands. For the three specimens with initial void ratios e_0 being 0.818, 0.746, and 0.662, respectively, the slopes of their compression curves in the e-lnp' coordinate system are essentially identical. The simulation results are in substantial agreement with the experimental results.

(1) Calculation Steps

For triaxial undrained shear, there are

$$d_t p' = \frac{1}{3} d_t \sigma_1 - d_t u_l \tag{6.31}$$

$$d_t q = d_t \sigma_1 \tag{6.32}$$

Therefore, it can be obtained that

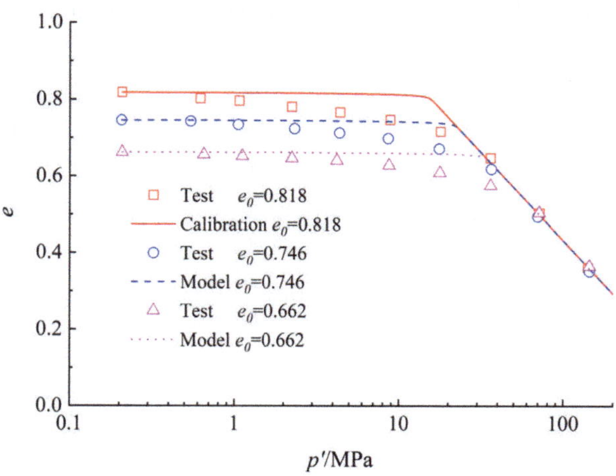

Fig. 6.10 Simulation of consolidation compression curve of saturated Ottawa sand

$$d_t p' = \frac{1}{3} d_t q - d_t u_l \tag{6.33}$$

According to Eqs. (6.8) and (6.9), the time derivatives of p' and q take the form:

$$d_t p' = \frac{\rho_l G_s p'}{\lambda \rho_d} d_t \varepsilon_v + B \begin{bmatrix} 8.75\left(\varepsilon_v^e\right)^{1.5} \\ +0.75\zeta\left(\varepsilon_v^e\right)^{-0.5}\left(\varepsilon_s^e\right)^2 \end{bmatrix} d_t \varepsilon_v^e + 3\zeta B\left(\varepsilon_v^e\right)^{0.5} \varepsilon_s^e d_t \varepsilon_s^e \tag{6.34}$$

$$d_t q = \frac{p_l G_s q}{\lambda \rho_d} d_t \varepsilon_v + 1.5\sqrt{6}\zeta B\left(\varepsilon_v^e\right)^{0.5} \varepsilon_s^e d_t \varepsilon_v^e + \sqrt{6}\zeta B\left(\varepsilon_v^e\right)^{1.5} d_t \varepsilon_s^e \tag{6.35}$$

For airy sandy soils, undrained shear will result in compressive deformation due to the internal air bubbles, i.e., cause a change in body strain, and therefore $d_t\varepsilon_v$ is not 0. From Eq. (6.30), the rate of change of body strain is related to the rate of change of pore pressure as follows:

$$d_t u_l = \frac{u + p_{atm}}{\left(1 - s_r + h s_r\right) n} d_t \varepsilon_v \tag{6.36}$$

Based on the above-mentioned equations and combined with the thermodynamic constitutive model of saturated sand proposed in Sect. 6.2, the undrained shear tests of gas-containing sands can be simulated and calculated.

(2) Undrained Shear Simulation of Saturated Ottawa Sands

Figures 6.11 and 6.12, respectively, show the simulation of the stress-strain curves and the effective stress paths during the undrained shearing of saturated specimens 11/12. The initial void ratios of the two specimens are basically the same, but the initial confining pressures are different. Both specimens exhibit shear softening, and the magnitudes of the peak shear stress and the shear stress at the critical

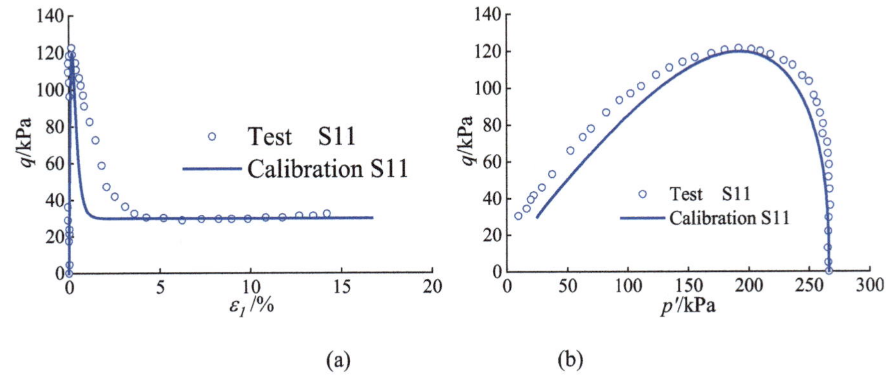

Fig. 6.11 Simulated results of undrained shear test of saturated sand sample S11, $e_0 = 0.89$, $p_0 = 266$ kPa: (**a**) stress-strain curve; (**b**) effective stress path

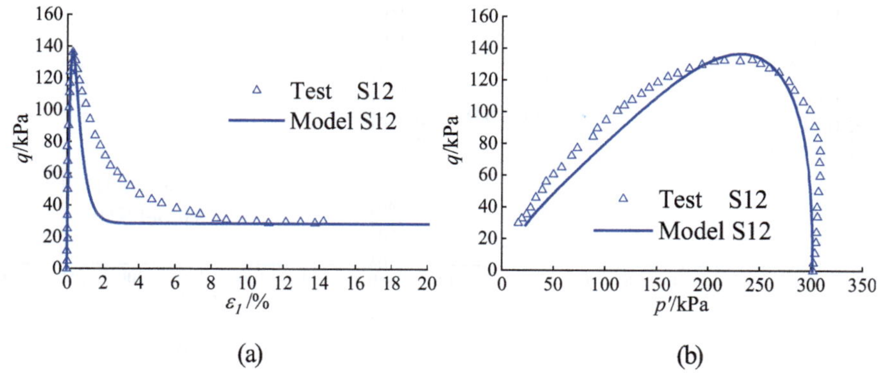

Fig. 6.12 Simulated results of undrained shear test of saturated sand sample S12, e_0 = 0.91, p_0 = 302 kPa: (**a**) stress-strain curve; (**b**) effective stress path

state are also basically the same. It can be seen that the softening rate of the simulated stress-strain curve is faster than that of the experimental result. The simulated results of the peak shear stress, the shear stress value at the critical state, and the effective stress path are all in agreement with the experimental results. Overall, the model can fairly well reflect the undrained shear process of saturated sands.

(3) Undrained Shear Simulation of Gas-containing Sands

In the dilatancy equation, an expression for the critical state void ratio e_c needs to be provided. Sasitharan et al. [34] presented the critical state line for Ottawa sands as:

$$e_c = 0.864 - 0.0168\log\left(p'\right) \tag{6.37}$$

Grozic et al. [17] presented the critical state line for gas-containing Ottawa sands based on Eq. (6.37) as:

$$e_c = 0.92 - 0.0159\log\left(p'\right) \tag{6.38}$$

Yang and Li [42] presented the critical state line for gas-containing sands as:

$$e_c = e_{c0} + \left(1 - s_{r0}\right)^{0.5} \Delta e_{c0} - \lambda_c \log\left(p'\right) \tag{6.39}$$

In this chapter, the critical state line of Ottawa sands presented by Grozic et al. [17] (Eq. (6.38)) is adopted for calculation. Figures 6.13 and 6.14, respectively, show the simulation results of Specimens 10 and 25, including the stress-strain curves, effective stress paths, changes in the void ratio and saturation. The saturations of the two specimens are both around 0.9, and the initial void ratios are 0.98 and 0.85, respectively. Specimen 10 exhibits strain softening. During the undrained shear process, due to the increase in pore-water pressure, the internal air bubbles are compressed, the void ratio decreases, and the saturation increases. The simulation

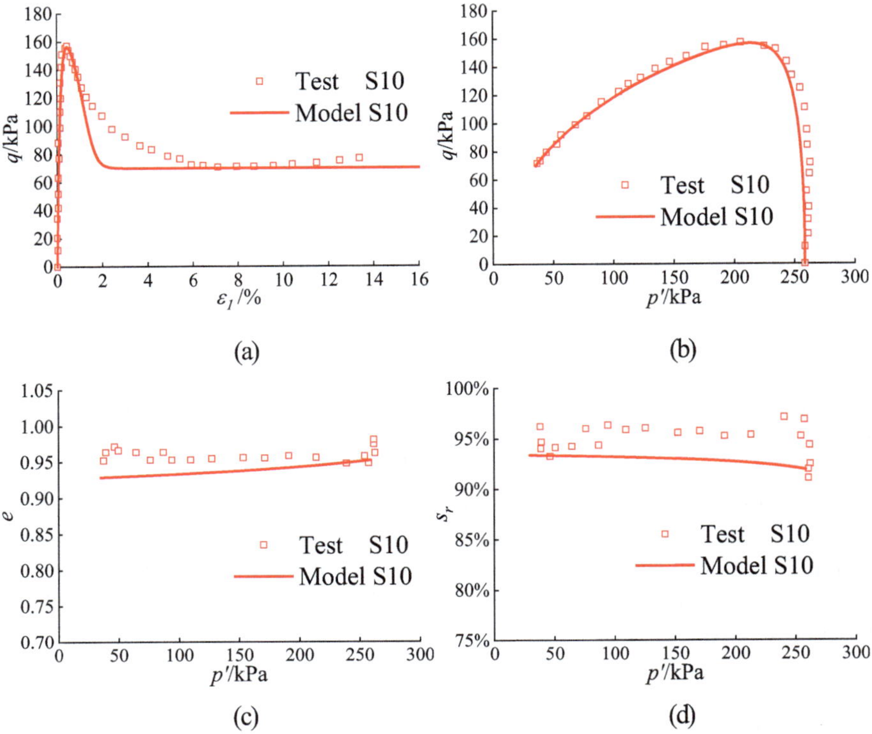

Fig. 6.13 Simulated results of undrained shear test of gassy sand sample S10, e_0 = 0.98, p_0 = 259 kPa, sr0 = 0.9: (**a**) stress-strain curve; (**b**) effective stress path; (**c**) variation of void ratio with effective stress; (**d**) variation of saturation with effective stress

results are in agreement with the experimental results. Specimen 25 experiences shear failure. After the shear stress reaches its peak, it softens and reaches a state where the shear stress is 0, and the effective stress also decreases to 0 at this time. The model can reflect this shear-failure phenomenon. Meanwhile, the void ratio decreases and the saturation increases. Although the amplitude of the simulation results is relatively larger than that of the experimental results, the overall change trend can be reflected.

Figure 6.15 presents the simulation results of the shear hardening of gas-containing sand Specimen 21. The saturation of the specimen is relatively low, at 0.8. That is to say, the internal gas content is relatively high, and the volumetric strain of the specimen is more likely to change, which will increase the drainage tendency of the specimen during undrained shearing. Judging from the experimental results, the shear stress has been showing an increasing trend with the shear strain. However, the simulation results show that after reaching a certain shear strain, the shear stress will tend to be stable. There are also certain differences between the simulation results of the effective stress path and the experimental results. During the initial shearing, due to the relatively large pore-gas content, it is

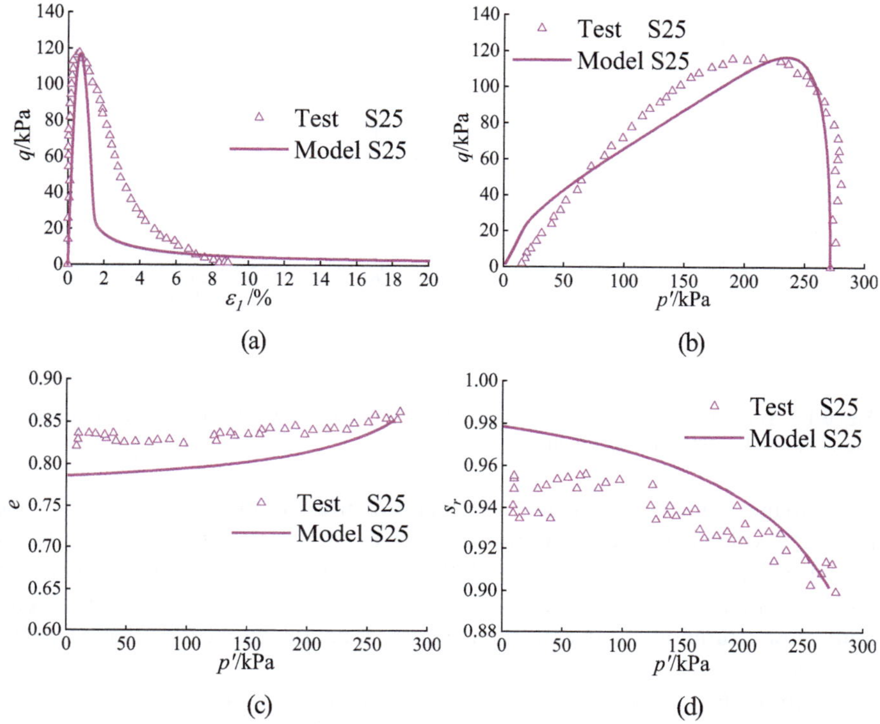

Fig. 6.14 Simulated results of undrained shear test of gassy sand sample S25, e_0 = 0.85, p_0 = 272 kPa, s_{r0} = 0.91: (**a**) stress-strain curve; (**b**) effective stress path; (**c**) variation of void ratio with effective stress; (**d**) variation of saturation with effective stress

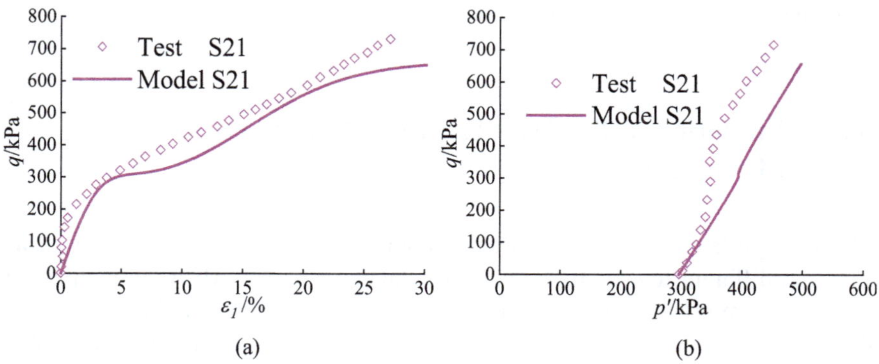

Fig. 6.15 Simulated results of undrained shear test of gassy sand sample S21, e_0 = 0.9, p_0 = 296 kPa, s_{r0} = 0.91: (**a**) stress-strain curve; (**b**) effective stress path

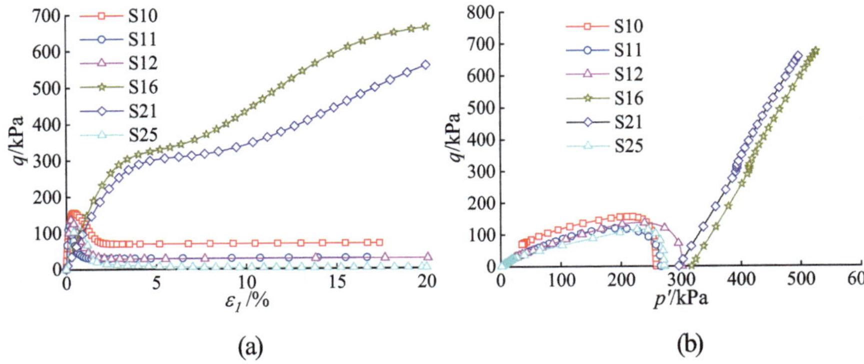

Fig. 6.16 Simulation of undrained shear test of sand samples with different saturation: (**a**) stress-strain curve; (**b**) effective stress path

closer to the drainage path, and then gradually approaches the critical state line. The degree to which the simulation results tend to the drainage path is relatively larger than that of the experimental results. Although there are certain deviations between the simulation results and the experimental results for gas-containing soil samples with a low saturation, the overall change trend of undrained shear hardening can be reflected.

Figure 6.16 presents a summary of the simulation results of the stress-strain curves and effective stress paths for the above-mentioned specimens and Specimen 16. The saturation of Specimen 16 is 0.816, which is similar to that of Specimen 21, and it also exhibits strain hardening. It can be observed that when the saturation is relatively low (less than or equal to 0.8), the gas-containing sand mainly shows strain hardening, while when the saturation is relatively high, it mainly shows strain softening and may even experience shear failure.

6.6 Simulation of the Temperature Characteristics of Gassy Sand

The ideal gas law states that for the same amount of gas, the value of $u_g V_g$ (pressure) varies at different temperatures. For gas-containing sands, when the temperature rises, the pore pressure also increases. After the temperature stabilizes, over time, the pore-water pressure gradually dissipates and eventually reaches a stable state. That is, an increase in temperature leads to the generation of excess pore-water pressure, and over time, the generated excess pore-water pressure also gradually dissipates. In this process, according to the assumed conditions, the pore-gas pressure undergoes the same change. According to the ideal gas law, when the pore pressure has completely dissipated, the gas volume will increase and has a proportional relationship with the temperature, that is, $V_1/V_0 = T_1/T_0$. Here, V_0 and T_0 are the gas volume and the reference temperature, respectively, and V_1 and T_1 are the gas volume and the temperature value after the temperature increase, respectively.

Gas-containing sands are generally assumed to have a continuous and incompressible liquid phase and a discontinuous and compressible gas phase. Since the gas bubbles are free in the pores of the soil skeleton, it can be assumed that they have no influence on the force and deformation of the soil skeleton. Therefore, the effective stress in gas-containing sands generally has the same form as that in saturated soils, that is, Terzaghi's effective stress principle is still applicable.

Based on the above-proposed thermodynamic constitutive model that can reflect the temperature characteristics of saturated sands, this model takes into account the energy dissipation at the particle level caused by temperature changes, and establishes a connection between the energy dissipation process of saturated sands and the macroscopic mechanical behavior through the migration coefficient and the energy density function model. At the same time, combining the dilatancy equation considering the state parameter with the model parameters can effectively describe the influence of temperature on the strength and deformation characteristics of saturated sands under different relative densities and effective confining pressures. Assuming that the thermal change of the gas phase has no effect on the deformation of the soil skeleton (the gas bubbles are still free in the pores of the soil skeleton), on this basis, combined with the above-given relationships, the influence of temperature on the mechanical properties of gas-containing sands is analyzed.

(1) Influence of Temperature on the Drained Shear Characteristics of Gas-containing Sands

Zhu et al. conducted experimental studies on the drained shear characteristics of gas-containing sands through an improved temperature-controlled triaxial test device, and found that the drained shear strength of gas-containing sands increases with the increase of temperature. Based on the characteristics of gas-containing sands, under drained conditions, the shear characteristics of gas-containing sands are similar to those of saturated sands. As the temperature increases, the drained shear strength of saturated sands also increases. Under the same temperature conditions, the drained shear strength of gas-containing sands is also higher than that of saturated sands under the same initial conditions. During the shearing process, the enclosed gas bubbles in the pore water of the gas-containing sands are forced to migrate and are trapped in one place due to capillary resistance. As the shearing progresses, more and more gas bubbles accumulate, and larger gas bubbles are formed around the particle contact parts, thereby generating suction and increasing the shear resistance, which in turn increases the shear strength. In addition, as the temperature increases, the gas bubbles expand due to heating, and the dissolved gas is released, resulting in an increase in the gas volume. Therefore, the increase in temperature enhances this shear characteristic of gas-containing sands (Fig. 6.17).

(2) Influence of Temperature on the Undrained Shear Characteristics of Gas-containing Sands

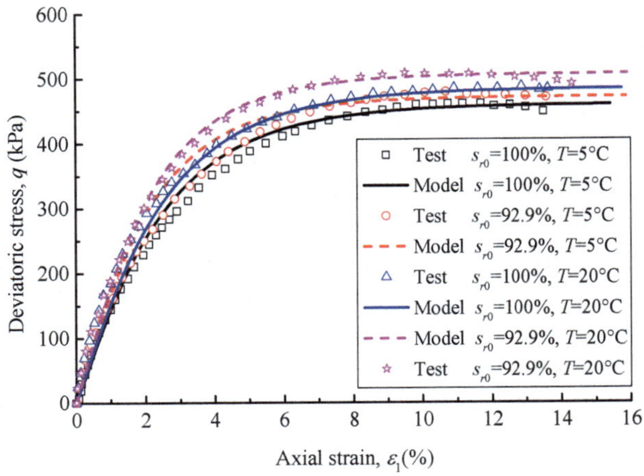

Fig. 6.17 The stress-strain curves of gassy sand specimens at different temperatures

Since gas-containing sands are generally found in marine sedimentary soils and are approximately in an undrained condition, more research has been focused on their undrained shear characteristics. Some scholars have studied the undrained shear characteristics of gas-containing sands under the action of ambient temperature and found that gas-containing sands exhibit strain softening or strain hardening during undrained shearing, depending on their initial state (loose or dense), the magnitude of the initial saturation and void ratio, etc. For loose gas-containing sands, when the initial saturation is relatively large (>90%), that is, the gas bubble content is relatively small, it generally shows strain softening and may even experience liquefaction. When the saturation is relatively small (80–90%), that is, the gas bubble content is relatively large, the gas-containing sand generally exhibits strain hardening. Under the same initial conditions, the undrained shear strength of gas-containing sands is slightly higher than that of saturated sands. For relatively dense sands, when they are completely water-saturated, they generally show shear hardening. When there are gas bubbles present, the shear strength decreases significantly, with a slight shear softening phenomenon. Moreover, when the gas content is higher, the corresponding shear strength is slightly lower. This indicates that for dense sands, the presence of gas bubbles weakens the undrained shear strength of the sands. Therefore, according to the experimental data, it can be considered that the gas bubble volume content has a promoting and enhancing effect on the shear strength of loose gas-containing sands, but has a weakening effect on dense sands.

For loose or dense saturated sands, the experimental data show that as the temperature increases, the undrained shear strength tends to decrease, as shown in Fig. 6.18). The test is carried out by first performing thermal consolidation and then

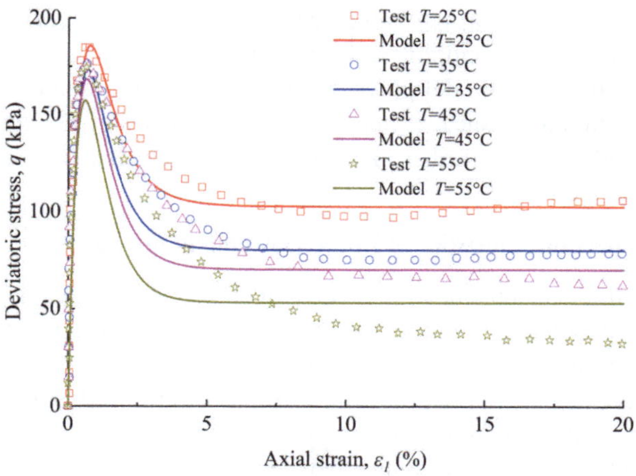

(a)

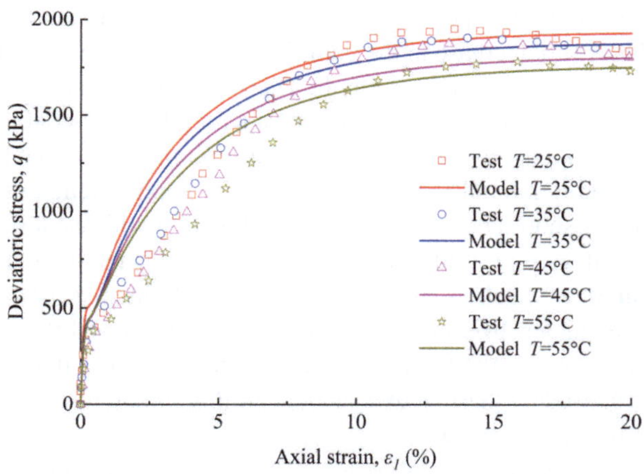

(b)

Fig. 6.18 Simulation of undrained shear stress-strain relationship of saturated sand samples with different relative compactness varying with temperature (p′ = 400 kpa, u = 300 kpa): (**a**) compactness dr = 20%; (**b**) compactness dr = 70%

undrained shearing. For both loose and dense specimens, as the temperature increases, the tested soil samples generate negative thermal volumetric strain (thermal expansion). Therefore, a soil sample at a higher temperature is equivalent to having a larger initial void ratio during shearing, that is, the soil sample is looser, so the undrained shear strength decreases slightly with the increase of temperature. The theoretical model proposed in this paper can capture this temperature characteristic of the sand.

In the existing literature, the author has not found any direct research on the temperature effect of the undrained shear characteristics of gas-containing sands. Therefore, based on the model given in this paper, only a qualitative simulation analysis of the undrained shear temperature characteristics of gas-containing sands has been carried out, and further analysis can be conducted in combination with relevant tests in the future.

(3) Temperature Characteristics of Loose Gas-containing Sands

During the thermal consolidation process of gas-containing sands, when the temperature is increased, excess hydrostatic pore pressure will be generated. Over time, the generated excess pore pressure will gradually dissipate, and after stabilization, the excess hydrostatic pore pressure will dissipate to zero. At this time, according to the previously mentioned assumed conditions, during the heating process, the pressure of the pore gas first increases and then decreases. Combining with the ideal gas equation, the volume of the pore gas will increase, which is equivalent to a decrease in saturation and is beneficial to the increase of the shear strength.

However, from another perspective, similar to saturated sands, an increase in temperature leads to a decrease in the undrained shear strength, which is not conducive to the increase of the shear strength. Therefore, the effects of the gas bubble volume and temperature on the undrained shear strength of gas-containing sands are in an offsetting relationship. The specific degrees of influence of the two are related to the initial conditions and the magnitude of the applied temperature load, and statistical analysis should be carried out in combination with the test data.

Based on the degree of influence of saturated sands by temperature changes and the experimental data such as the changes in the undrained shear strength of gas-containing sands under different saturations, it can be inferred that the increase in shear strength caused by the increase in gas volume is greater than the decrease in shear strength caused by thermal expansion, and this should be more obvious when the saturation is relatively small.

For loose gas-containing sands, according to the simulation and prediction results of the proposed theoretical model, it can be seen that with the increase in temperature, the peak strength and the final critical state of the undrained shear of loose gas-containing sands will increase slightly, which is consistent with the above-inferred law. Figure 6.19 shows the simulation results of strain softening when the saturation is relatively high ($Sr = 0.98$). It can be seen that as the temperature increases, the difference between the peak strength and the final critical state strength becomes smaller and smaller, and the softening phenomenon is weakened. Figure 6.20 shows the simulation results of strain hardening when the saturation is relatively small ($Sr = 0.81$). As the temperature increases, the hardening phenomenon is also enhanced.

However, whether it is softening or hardening, the amplitude of the increase in the undrained shear strength is relatively small, which is the result of the mutual

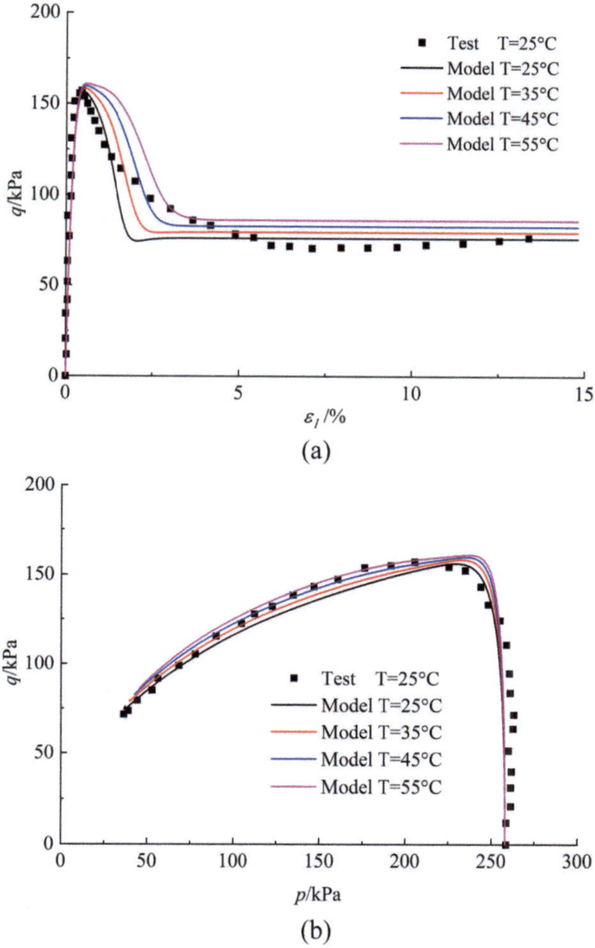

Fig. 6.19 Simulated results of undrained shear test of gassy sand sample 10, e0 = 0.98, p0 = 259 kPa, sr0 = 0.9 (test data after [4]): (**a**) stress-strain curve; (**b**) effective stress path

offset of the influence degrees of temperature and gas bubble volume as mentioned above.

Similarly, He and Chu also conducted experimental studies on the undrained shear characteristics of gas-containing sands with different initial saturations. They also found that as the saturation decreases and the volume of the gas phase increases, the undrained shear strength of the gas-containing sands increases. The model prediction results show that when the temperature increases, the shear strength of the gas-containing sands also increases slightly, which is consistent with the conclusion drawn previously, as shown in Fig. 6.21.

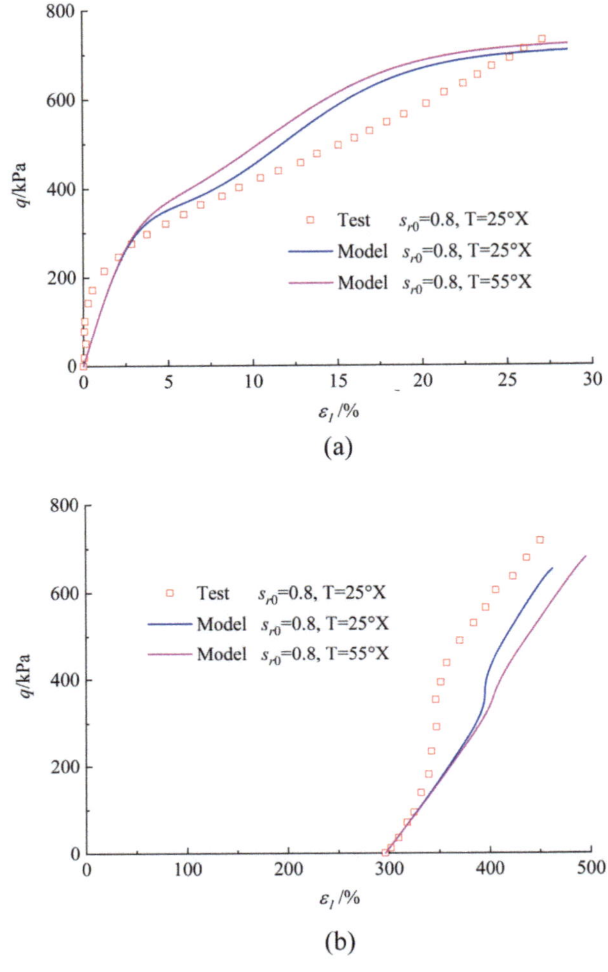

Fig. 6.20 Simulation and prediction of undrained shear properties of loose aerated sand under different temperatures sr0 = 0.81: (**a**) stress-strain curve; (**b**) effective stress path

(4) Temperature Characteristics of Dense Gas-containing Sands

For relatively dense sands, studies have found that the presence of gas bubbles will reduce the undrained shear strength of the sands. As shown in Fig. 6.22, Rad et al. [39] conducted undrained shear tests on dense gas-containing sands (with a relative density Dr. of about 85%). It was found that when the sands are completely water-saturated (Sr = 1), (α = 0), where α represents the amount of dissolved gas, and (α = 0) indicates that there is no dissolved gas), the undrained shear strength of the dense sands is very high, and the negative pore pressure generated by dilatancy is also relatively large. When there is dissolved gas in the pore water (Sr = 1, α = 100%), the undrained shear strength decreases significantly, and a slight shear softening occurs. When there is free gas (Sr = 0.9, α = 100%), the undrained shear

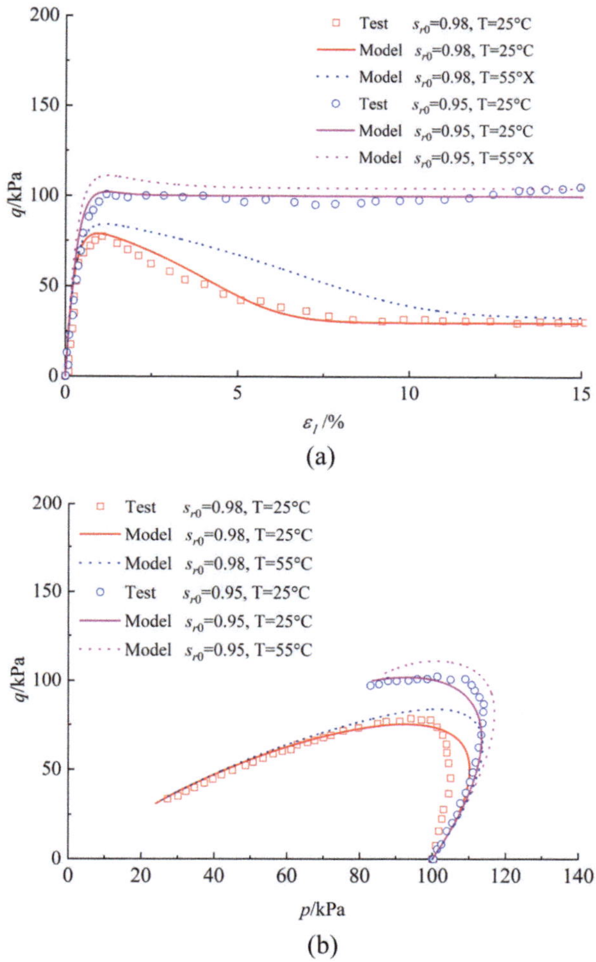

Fig. 6.21 Simulation of temperature characteristics of aerated sand with different initial saturation: (**a**) stress-strain curve; (**b**) effective stress path

strength decreases further. This indicates that the presence of gas reduces the undrained shear strength of the dense sands, having an "adverse" effect. From the above analysis, it can be seen that an increase in temperature will also reduce the undrained shear characteristics of dense sands. At this time, temperature and gas bubbles have the same effect. Therefore, increasing the temperature will decrease the undrained shear strength of dense gas-containing sands, and the corresponding negative pore pressure will also decrease.

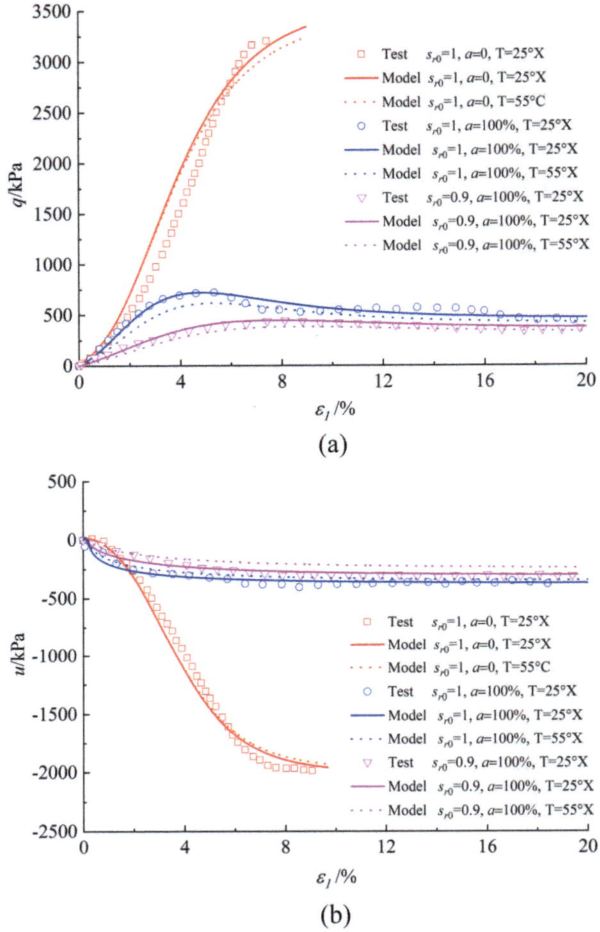

Fig. 6.22 Comparison between the measured and predicted excess pore water pressure of dense gassy Baskarp sand of different saturation in undrained triaxial compression (Data from [39])

6.7 Summary

This chapter, based on the theoretical model proposed in Chap. 2, which is grounded in the theory of granular matter hydrodynamics, and in combination with the dilatancy equation incorporating the state parameter, presents a thermodynamic constitutive model capable of describing the dilatancy of saturated sands. The model introduces the concepts of granular entropy and granular temperature to depict the irreversible deformation within the sands. It establishes a connection between the dissipation mechanism of saturated sands and the macroscopic mechanical behavior through the migration coefficient and the energy density function. The model has a relatively simple form, and with a set of model parameters, it can describe the influence of changes in relative density and effective confining pressure on the strength and deformation characteristics of saturated sands during the shearing process. Through the comparison between the simulation calculations and the results of the isotropic compression, triaxial undrained, and drained shear tests on Toyoura sand, the model's ability to describe the compression and shear characteristics of saturated sands has been verified.

Regarding gas-containing sands, it is considered that the presence of gas bubbles only increases the compressibility and does not affect the deformation of the soil skeleton. Based on the presented thermodynamic model of saturated sands, and in combination with the relationship established by Hilf [37] between the pore gas pressure and the change in pore gas volume, a thermodynamic constitutive model for gas-containing sands has been developed. It has been found that gas-containing sand specimens exhibit shear hardening or shear softening phenomena in triaxial undrained shear tests, depending on the magnitudes of the initial saturation and void ratio. The presence of gas has the effect of reducing rather than eliminating the sensitivity of loose sands to flow liquefaction. The triaxial undrained shear tests conducted by Grozic et al. [4] on remolded Ottawa sands with different saturations have verified the model's ability to describe the undrained shear characteristics of gas-containing sands.

Under different conditions, the influence of temperature on the shear characteristics of gas-containing sands also varies. Under drained conditions, when the temperature rises, the corresponding shear strength also increases, which has been verified by relevant tests. This is related to the increase in gas volume caused by the temperature rise. For the undrained situation, when the gas-containing sands are relatively loose, the temperature and the gas volume show an offsetting effect, and the influence degree of temperature is greater. Therefore, the increase in temperature can, to a certain extent, increase the shear strength. When the gas-containing sands are relatively dense, both the temperature and the gas volume have the effect of reducing the shear strength. Thus, the increase in temperature decreases the shear strength, and the larger the gas volume, the more obvious this phenomenon becomes.

References

1. Li XS, Dafalias YF, Wang ZL (1999) State-dependant dilatancy in criticalstate constitutive modelling of sand. Can Geotech J 36(4):599–611
2. Christian HA, Cranston RE (1997) A methodology for detecting free gas in marine sediments. Can Geotech J 34(2):293–304
3. Amaratunga A, Grozie JL (2009) On the undrained unloading behaviour of gassy sands. Can Geotech J 46(11):1267–1276
4. Grozic JLH, Robertson PK, Morgestern NR (1999) The behavior of loose gassy sand. Can Geotech J 36(3):482–492
5. Fourie AB, Hofmann BA, Mikula RJ, Lord ERF, Robertson PK (2001) Partially saturated tailings sand below the phreatic surface. Geotechnique 51(7):577–585
6. Yoshimi Y, Tanaka K, Tokimatsu K (1989) Liquefaction resistance of a partially saturated sand. Soils Found 29(3):157–162

7. Xia H, Hu T (1991) Effects of saturation and back pressure on sand liquefaction. J Geotech Eng 117(9):1347–1362
8. Tamura S, Tokimatsu K, Abe A, Sato M (2002) Effects of air bubbles on B-value and P-wave velocity of a partially saturated sand. Soils Found 42(1):121–129
9. Tsukamoto Y, Ishihara K, Nakazawa H, Kamada K, Huang Y (2002) Resistance of partly saturated sand to liquefaction with reference to longitudinal and shear wave velocities. Soils Found 42(6):93–104
10. Pietruszczak S, Pande GN, Oulapour M (2003) A hypothesis for mitigation of risk of liquefaction. Geotechnique 53(9):833–838
11. Altun S, Goktepe AB (2006) Cyclic stress-strain behavior of partially saturated soils. Unsat Soil:497–507
12. Bouferra R, Benseddiq N, Shahrour I (2007) Saturation and preloading effects on the cyclic behavior of sand. Int J Geomech 7(5):396–401
13. Vega-Posada CA, Finno RJ, Zapata-Medina DG (2014) Effect of gas on the mechanical behavior of medium-dense sands. J Geotech Geoenviron 140(11):04014063
14. He J, Chu J (2014) Undrained responses of microbially desaturated sand under monotonic loading. J Geotech Geoenviron Eng 140(5):04014003
15. He J, Chu J, Liu H (2014) Undrained shear strength of desaturated loose sand under monotonic shearing. Soils Found 54(4):910–916
16. Pietruszczak S, Pande GN (1996) Constitutive relations for partially saturated soils containing gas inclusions. J Geotech Eng 122(1):50–59
17. Grozic JLH, Imam SMR, Robertson PK, Morgenstern NR (2005) Constitutive modeling of gassy sand behaviour. Can Geotech J 42:812–829
18. Liu K, Xue JF, Yang M (2016) Deformation behaviour of geotechnical materials with gas bubbles and time dependent compressible organic matter. Eng Geol 213:98–106
19. Lü XL, Huang MS, Andrade JE (2018) Modeling the static liquefaction of unsaturated sand containing gas bubbles. Soils Found 58(1):122–133
20. Roscoe KH, Scgofield AN, Wroth CP (1958) On the yielding of soils. Géotechnique 8(1):22–53
21. Poorooshasb HB, Pietruszczak S (1986) A generalized flow theory for sand. Soils Found 26(2):1–15
22. Wang Z L，Dafalias Y F，Shen C K. Bounding surface hypoplasticity model for sand. J Eng Mech 1990, 116(5): 983–1001
23. Li XS, Dafalias YF, Wang ZL (1999) State-dependant dilatancy in critical state constitutive modelling of sand. Can Geotech J 36(4):599–611
24. Been K, Jefferies MG (1985) A state parameter for sands. Géotechnique 35(2):99–112
25. Jefferies MG (1993) Nor-sand: a simple critical state model of sand. Géotechnique 43(1):91–103
26. Manzari MT, Dafalias YF (1997) A critical state two-surface plasticity model for sands. Géotechnique 47(2):255–272
27. Li XS, Wang Y (1998) Linear representation of steady-state line for sand. J Geotech Geoenviron Eng 124(12):1215–1217
28. Wan RG, Guo RG (1998) A simple constitutive model for granular solis: modified stress-dilatancy approach. Comput Geotech 22(2):109–113
29. Li XS, Dafalias YF (2000) Dilatancy for cohesionless soils. Geotechnique 50(4):449–460
30. Li XS (2002) A sand model with state-dependent dilatancy. Géotechnique 52(3):173–186
31. Zhang ZC, Cheng XH (2016) A thermo-mechanical coupled constitutive model for clay based on extended granular solid hydrodynamics. Comput Geotech 80:373–382
32. Jiang YM, Liu M (2009) Granular solid hydrodynamics. Granul Matter 11(3):139–156
33. Schofield A N. Disturbed soil properties and geotechnical design London: Thomas Telford Publishing, 2005
34. Sasitharan S, Robertson PK, Sego DC, Morgenstern NR (1994) State-boundary surface for very loose sand and its practical implications. Can Gcotech J 31:321–331

35. Mesri G, Vardhanabhuti B (2009) Compression of granular materials. Can Geotech J 46:369–392
36. Verdugo R, Ishihara K (1996) The steady state of sandy soils. Soils Found 36(2):81–91
37. Hilf JW (1948) Estimating construction pore pressures in rolled earth dams. In: Proceedings of the 2nd international conference on soil mechanics and foundation engineering, Rotterdam, Netherlands, vol 3, pp 234–240
38. Finno RJ, Zhang Y, Buscarnera G (2017) Experimental validation of Terzaghi's effective stress principle for gassy sand. J Geotech Geoenviron Eng 143(12):04017092
39. Rad NS, Vianna AJD, Berre T (1994) Gas in soils. II: effect of gas on undrained static and cyclic strength of sand. J Geotech Eng 120(4):716–737
40. Fredlund DG, Rahardjo H (1993) Soil mechanics for unsaturated soils. Wiley, New York
41. Pestana JM, Whittle AJ (1995) Compression model for cohesionless soils. Géotechnique 45(4):611–631
42. Yang J, Li XS (2004) State-dependent strength of sands from the perspective of unified modeling. J Geotech Geoenviron Eng 130(2):186–198

Chapter 7
Conclusion

Based on the theory of thermodynamics of granular materials, a multi-field coupled constitutive model framework for unsaturated soils was established. Under this model framework, the application analysis of the model was carried out, respectively, for the cases of saturated clays, fine-grained gassy soil, saturated sand, and gassy sand. The main research achievements and conclusions are as follows:

1. Based on the theory of thermodynamics of granular materials and the mixture theory, combined with the improved SWCC model, and considering the energy dissipation processes at the granular level caused by changes in temperature and saturation, a multi-field coupled constitutive model for unsaturated soils was proposed. Different from general unsaturated soil models, this model does not involve concepts such as yield criteria, flow rules, and hardening criteria. Instead, it introduces the concepts of granular entropy and granular temperature, theoretically determines the composition of the dissipative structure of the thermodynamic system and the relationship of the migration coefficients, and describes the dissipative mechanism of the soil and its induced macroscopic physical and mechanical behaviors as a model of migration coefficients and energy functions. By combining the total differential equation of thermodynamics for unsaturated soils, the conservation equation, and the entropy increase equation, a thermodynamic identity was established, and the constitutive relationship of inelastic deformation was derived through the thermodynamic identity and the dissipative relationship. These enable the model to more deeply describe the energy dissipation mechanism of unsaturated soils and the resulting macroscopic mechanical properties, thus enabling the establishment of a multi-field coupled model for unsaturated soils in complex environments within a unified theoretical framework. Through comparison with existing experimental results, the ability of the model to describe the multi-field coupling characteristics of unsaturated soils was demonstrated.

G. Yang, *Multi-field Coupled Constitutive Model for Soils Based on Granular Thermodynamics Theory*, https://doi.org/10.1007/978-981-95-0249-3_7

2. On the basis of the proposed general framework of the multi-field coupled model for unsaturated soils, for the special case of saturated clays, the model was degenerated to the level of saturated soils, and a thermodynamic constitutive model capable of describing the thermo-hydro-mechanical coupling characteristics of saturated clays was established. Through the comparative analysis between the results of thermal consolidation tests and the experimental results of the thermal response of saturated silty clays under cyclic temperature loads and the model simulation results, the ability of the proposed model to describe the thermal consolidation characteristics of saturated clays was verified.
3. Based on the theoretical framework of the proposed multi-field coupled model for unsaturated soils in this paper, combined with the equation of state of an ideal gas and considering the influence of the gas phase on the plastic deformation of the soil skeleton, a thermodynamic constitutive model capable of describing the mechanical properties and temperature effects of gas-containing fine-grained soils was developed, and the effective stress of gas-containing fine-grained soils was analyzed. According to the concept of the large gas bubble model proposed by Wheeler [1], gas-containing fine-grained soils were regarded as composed of large gas bubbles and the surrounding saturated soil matrix. In the model, the gas pressure is affected by the total stress and the pore water pressure and has a linear relationship with the total stress, while the saturated soil matrix part is controlled by the effective stress-like parameter. Through the comparison between the simulation results and the experimental results, the ability of the model to describe the consolidation compression and undrained shear characteristics of gas-containing fine-grained soils was verified. In addition, the temperature characteristics of gas-containing fine-grained soils were discussed, and using the proposed model, the temperature effects of gas-containing fine-grained soils under different drainage conditions and over-consolidation ratios were analyzed and simulated, including the influence of temperature on compressibility, thermal shrinkage, pore water pressure, and shear characteristics.
4. Based on the theoretical framework of the proposed multi-field coupled constitutive model for unsaturated soils in this paper, for the case of saturated sands, a dilatancy equation considering the state parameter was introduced, thus forming a thermodynamic constitutive model capable of describing the dilatancy of saturated sands. The model has a relatively simple form, and with a set of model parameters, it can describe the influence of changes in relative density and effective confining pressure on the strength and deformation characteristics of saturated sands during the shearing process. Based on the comparison between the simulation calculation results and the experimental results, the ability of the model to describe the compression and shear characteristics of saturated sands was verified.
5. Based on the thermodynamic model of saturated sands, combined with the relationship established by Hilf [2] between the pore gas pressure and the change in pore gas volume, a thermodynamic constitutive model for gas-containing sands was developed. It was found that for gas-containing sands, the presence of gas bubbles was considered only to increase the compressibility and would not affect

the deformation of the soil skeleton. Gas-containing sand specimens exhibited shear hardening or shear softening phenomena in triaxial undrained shear tests, depending on the magnitudes of the initial saturation and void ratio. The presence of gas has the effect of reducing rather than eliminating the sensitivity of loose sands to flow liquefaction. Through the triaxial undrained shear tests conducted by Grozic et al. [3] on remolded Ottawa sands with different saturations, the ability of the model to describe the undrained shear characteristics of gas-containing sands was verified. For drained shear, an increase in temperature will increase the shear strength of gas-containing sands, and the simulation results can be in good agreement with the experimental data. For undrained shear, an increase in temperature will increase the undrained shear strength of loose gas-containing sands, while decreasing the undrained shear strength of dense gas-containing sands.

References

1. Wheeler SJ (1988) A conceptual model for soils containing large gas bubbles [J]. Géotechnique 38(3):389–397
2. Hilf JW (1948) Estimating construction pore pressures in rolled earth dams. In: Proceedings of the 2nd international conference on soil mechanics and foundation engineering, Rotterdam, Netherlands, vol 3, pp 234–240
3. Grozic JLH, Robertson PK, Morgestern NR (1999) The behavior of loose gassy sand [J]. Can Geotech J 36(3):482–492

MIX
Papier aus verantwortungsvollen Quellen
Paper from responsible sources
FSC® C105338

If you have any concerns about our products,
you can contact us on
ProductSafety@springernature.com

In case Publisher is established outside the EU,
the EU authorized representative is:
Springer Nature Customer Service Center GmbH
Europaplatz 3, 69115 Heidelberg, Germany

Printed by Libri Plureos GmbH
in Hamburg, Germany